Machine Learning in Image Steganalysis

Machine Learning in Image Steganalysis

Hans Georg Schaathun

Ålesund University College, Norway

IEEE PRESS

A John Wiley & Sons, Ltd., Publication

Library of Congress CataloginginPublication Data

Schaathun, Hans Georg.
  Machine learning in image steganalysis / Hans Georg Schaathun.
     p. cm.
    Includes bibliographical references and index.
  ISBN 978-0-470-66305-9 (hardback)
  1. Machine learning. 2. Wavelets (Mathematics) 3. Data encryption (Computer science) I. Title.
  Q325.5.S285 2012
  006.3′1 – dc23

                                    2012016642

A catalogue record for this book is available from the British Library.

Print ISBN: 9780470663059

Typeset in 10/12.5 Palatino by Laserwords Private Limited, Chennai, India
Printed and bound in Singapore by Markono Print Media Pte Ltd

Contents

Preface

Books are conceived with pleasure and completed with pain. Sometimes the result is not quite the book which was planned.

This is also the case for this book. I have learnt a lot from writing this book, and I have discovered a lot which I originally hoped to include, but which simply could not be learnt from the literature and will require a substantial future research effort. If the reader learns half as much as the author, the book will have been a great success.

The book was written largely as a tutorial for new researchers in the field, be they postgraduate research students or people who have earned their spurs in a different field. Therefore, brief introductions to much of the basic theory have been included. However, the book should also serve as a survey of the recent research in steganalysis. The survey is not complete, but since there has been no other book[1] dedicated to steganalysis, and very few longer survey papers, I hope it will be useful for many.

I would like to take this opportunity to thank my many colleagues in the Computing Department at the University of Surrey, UK and the Engineering departments at Ålesund University College, Norway for all the inspiring discussions. Joint research with Dr Johann Briffa, Dr Stephan Wesemeyer and Mr (now Dr) Ainuddin Abdul Wahab has been essential to reach this point, and I owe them a lot. Most of all, thanks to my wife and son for their patience over the last couple of months. My wife has also been helpful with some of the proofreading and some of the sample photographs.

---

[1] We have later learnt that two other books appeared while this was being written: Rainer Böhme: *Advanced Statistical Steganalysis* and Mahendra Kumar: *Steganography and Steganalysis in JPEG images*.

Part I
Overview

1

Introduction

Steganography is the art of communicating a secret message, from Alice to Bob, in such a way that Alice's evil sister Eve cannot even tell that a secret message exists. This is (typically) done by hiding the secret message within a non-sensitive one, and Eve should believe that the non-sensitive message she can see is all there is. Steganalysis, on the contrary, is Eve's task of detecting the presence of a secret message when Alice and Bob employ steganography.

1.1 Real Threat or Hype?

A frequently asked question is *Who needs steganalysis?* Closely related is the question of who is using steganography. Unfortunately, satisfactory answers to these questions are harder to find.

A standard claim in the literature is that terrorist organisations use steganography to plan their operations. This claim seems to be founded on a report in *USA Today*, by Kelley (2001), where it was claimed that Osama bin Laden was using the Internet in an 'e-*jihad*' half a year before he became world famous in September 2001. The idea of the application is simple. Steganography, potentially, makes it possible to hide detailed plans with maps and photographs of targets within images, which can be left on public sites like e-Bay or Facebook as a kind of electronic *dead-drop*.

The report in *USA Today* was based on unnamed sources in US law enforcement, and there has been no other evidence in the public domain that terrorist organisations really are using steganography to plan their activities. Goth (2005) described it as a hype, and predicted that the funding opportunities enjoyed by the steganography community in the early years of the millennium would fade. It rather seems that he was correct. At least the EU and European research councils have shown little interest in the topic.

Machine Learning in Image Steganalysis, First Edition. Hans Georg Schaathun.
© 2012 John Wiley & Sons, Ltd. Published 2012 by John Wiley & Sons, Ltd.

Steganography has several problems which may make it unattractive for criminal users. Bagnall (2003) (quoted in Goth (2005)) points out that the acquisition, possession and distribution of tools and knowledge necessary to use steganography in itself establishes a traceable link which may arouse as much suspicion as an encrypted message. Establishing the infrastructure to use steganography securely, and keeping it secret during construction, is not going to be an easy exercise.

More recently, an unknown author in *The Technical Mujahedin* (Givner-Forbes, 2007; Unknown, 2007) has advocated the use of steganography in the *jihad*, giving some examples of software to avoid and approaches to evaluating algorithms for use. There is no doubt that the technology has some potential for groups with sufficient resources to use it well.

In June 2010 we heard of ten persons (alleged Russian agents) being arrested in the USA, and according to the news report the investigation turned up evidence of the use of steganography. It is too early to say if these charges will give steganalysis research a new push. Adee (2010) suggests that the spies may have been thwarted by old technology, using very old and easily detectable stego-systems. However, we do not know if the investigators first identified the use of steganography by means of steganalysis, or if they found the steganographic software used on the suspects' computers first.

So currently, as members of the general public and as academic researchers, we are unable to tell whether steganography is a significant threat or mainly a brain exercise for academics. We have no strong evidence of significant use, but then we also know that MI5 and MI6, and other secret services, who would be the first to know if such evidence existed, would hardly tell us about it. In contrast, by developing public knowledge about the technology, we make it harder for criminal elements to use it successfully for their own purposes.

1.2 Artificial Intelligence and Learning

Most of the current steganalysis techniques are based on machine learning in one form or another. Machine learning is an area well worth learning, because of its wide applications within medical image analysis, robotics, information retrieval, computational linguistics, forensics, automation and control, etc. The underlying idea is simple; if a task is too complex for a human being to learn, let's train a machine to do it instead. At a philosophical level it is harder. What, after all, do we really mean by *learning*?

Learning is an aspect of intelligence, which is often defined as the ability to learn. Machine learning thus depends on some kind of artificial intelligence (AI). As a scientific discipline machine learning is counted as a sub-area of AI, which is a more well-known idea at least for the general public. In contrast, our impression of what AI is may be shaped as much by science fiction as by science. Many of us would first think of the sentient computers and robots in the 1960s and 1970s literature, such as Isaac Asimov's famous robots who could only be kept from world domination by the three robotic laws deeply embedded in their circuitry.

As often as a dream, AI has been portrayed as a nightmare. Watching films like *Terminator* and *The Matrix*, maybe we should be glad that scientists have not yet managed to realise the dream of AI. Discussing AI as a scientific discipline today, it may be more fruitful to discuss the different sub-disciplines. The intelligent and sentient computer remains science fiction, but various AI-related properties have been realised with great success and valuable applications. Machine learning is one of these sub-disciplines.

The task in steganalysis is to take an *object* (communication) and classify this into one out of two classes, either the class of steganograms or the class of clean messages. This type of problem, of designing an algorithm to map objects to classes, is known as *pattern recognition* or classification in the literature. Once upon a time pattern recognition was primarily based on statistics, and the approach was analytic, aiming to design a statistical model to predict the class. Unfortunately, in many applications, the problem is too complex to make this approach feasible. Machine learning provides an alternative to the analytic approach.

A learning classifier builds a statistical model to solve the classification problem, by brute-force study of a large number of statistics (so-called *features*) from a set of objects selected for training. Thus, we say that the classifier learns from the study of the training set, and the acquired learning can later be used to classify previously unseen objects. Contrary to the analytic models of statistical approaches, the model produced by machine learning does not have to be comprehensible for human users, as it is primarily for machine processing. Thus, more complex and difficult problems can be solved more accurately.

With respect to machine learning, the primary objective of this book is to provide a tutorial to allow a reader with primary interest in steganography and steganalysis to use black box learning algorithms in steganalysis. However, we will also dig a little bit deeper into the theory, to inspire some readers to carry some of their experience into other areas of research at a later stage.

1.3 How to Read this Book

There are no 'don't do this at home' clauses in this book. Quite the contrary. The body of experimental data in the literature is still very limited, and we have not been able to run enough experiments to give you more than anecdotal evidence in this book. To choose the most promising methods for steganalysis, the reader will have to make his own comparisons with his own images. Therefore, the advice must be 'don't trust me; try it yourself'. As an aid to this, the software used in this book can be found at: **http://www.ifs.schaathun.net/pysteg/**.

For this reason, the primary purpose of this book has been to provide a hands-on tutorial with sufficient detail to allow the reader to reproduce examples. At the same time, we aim to establish the links to theory, to make the connection to relevant areas of research as smooth as possible. In particular we spend time on statistical methods, to explain the limitations of the experimental paradigms and understand exactly how far the experimental results can be trusted.

In this first part of the book, we will give the background and context of steganalysis (Chapter 2), and a quick introduction and tutorial (Chapter 3) to provide a test platform for the next part.

Part II is devoted entirely to feature vectors for steganalysis. We have aimed for a broad survey of available features, but it is surely not complete. The best we can hope for is that it is more complete than any previous one. The primary target audience is research students and young researchers entering the area of steganalysis, but we hope that more experienced researchers will also find some topics of interest.

Part III investigates both the theory and methodology, and the context and challenges of steganalysis in more detail. More diverse than the previous parts, this will both introduce various classifier algorithms and take a critical view on the experimental methodology and applications in steganalysis. The classification algorithms introduced in Chapters 11 and 12 are intended to give an easy introduction to the wider area of machine learning. The discussions of statistics and experimental methods (Chapter 10), as well as applications and practical issues in steganalysis (Chapter 14) have been written to promote thorough and theoretically founded evaluation of steganalytic methods. With no intention of replacing any book on machine learning or statistics, we hope to inspire the reader to read more.

2

Steganography and Steganalysis

Secret writing has fascinated mankind for several millennia, and it has been studied for many different reasons and motivations. The military and political purposes are obvious. The ability to convey secret messages to allies and own units without revealing them to the enemy is obviously of critical importance for any ruler. Equally important are the applications in mysticism. Literacy, in its infancy, was a privilege of the elite. Some cultures would hold the ability to create or understand written messages as a sacred gift. Secret writing, further obscuring the written word, would further elevate the inner-most circle of the elite. Evidence of this can be seen both in hieroglyphic texts in Egypt and Norse runes on the Scottish islands.

2.1 Cryptography versus Steganography

The term *steganography* was first coined by an occultist, namely Trithemius (c. 1500) (see also Fridrich (2009)). Over three volumes he discussed methods for encoding messages, occult powers and communication with spirits. The word 'steganography' is derived from the Greek words στεγανος (steganos) for 'roof' or 'covered' and γραφειν (grafein) 'to write'. Covered, here, means that the message should be concealed in such a way that the uninitiated cannot tell that there is a secret message at all. Thus the very existence of the secret message is kept a secret, and the observer should think that only a mundane, innocent and non-confidential message is transmitted.

Use of steganography predates the term. A classic example was reported by Herodotus (440 BC, Book V). Histiæus of Miletus (late 6th century BC) wanted to send a message to his nephew, Aristagoras, to order a revolt. With all the roads being guarded, an encrypted message would surely be intercepted and blocked.

Machine Learning in Image Steganalysis, First Edition. Hans Georg Schaathun.
© 2012 John Wiley & Sons, Ltd. Published 2012 by John Wiley & Sons, Ltd.

Even though the encryption would keep the contents safe from enemy intelligence, it would be of no use without reaching the intended recipient. Instead, he took a trusted slave, shaved his head, and tattooed the secret message onto it. Then he let the hair regrow to cover the message before the slave was dispatched.

Throughout history, techniques for secret writing have most often been referred to as *cryptography*, from χρυπτος (cryptos) meaning 'hidden' and γραφειν (grafein, 'to write'). The meaning of the word has changed over time, and today, the cryptography and steganography communities would tend to use it differently.

Cryptography, as an area of research, now covers a wide range of security-related problems, including secret communications, message authentication, identification protocols, entity authentication, etc. Some of these problems tended to have very simple solutions in the pre-digital era. Cryptography has evolved to provide solutions to these problems in the digital domain. For instance, historically, signatures and seals have been placed on paper copies to guarantee their authenticity. Cryptography has given us digital signatures, providing, in principle, similar protection to digital files. From being an occult discipline in Trithemius' time, it has developed into an area of mathematics in the 20th century. Modern cryptography is always based on formal methods and formal proofs of security, giving highly trusted modules to be used in the design of secure systems.

The most well-known problem in cryptography is obviously that of *encryption* or ciphers, where a plain text message is transformed into incomprehensible, random-looking data, and knowledge of a secret key is necessary to reverse the transformation. This provides secret communications, in the sense that the contents of the message are kept secret. However, the communication is not 'covered' in the sense of steganography. Communicating random-looking data is suspicious and an adversary will assume that it is an encrypted message.

It is not uncommon to see 'cryptography' used to refer to encryption only. In this book, however, we will consistently use the two terms in the way they are used by the cryptography community and as described above. Most of the current research in steganography lacks the formal and mathematical foundations of cryptography, and we are not aware of any practical steganography solutions with formal and quantitative proofs of security. Yet, as we shall see below, steganography can be cast, and has indeed been cast, as a cryptographic problem. It just seems too difficult to be *solved* within the strict formal framework of modern cryptography.

2.2 Steganography

Much has been written about steganography during the last 15–30 years. The steganography community derives largely from the signal- and image-processing community. The less frequent contributions from the cryptography and information theory communities do not always use the same terminology, and this can make it hard to see the connections between different works.

2.2.1 The Prisoners' Problem

The modern interest in steganography is largely due to Simmons (1983) who defined and explained a steganographic security model as a cryptographic problem, and presented it at the Number 1 international conference on cryptography, Crypto 1983.

This security model is known as *the prisoners' problem*. Two prisoners, Alice and Bob, are allowed to exchange messages, but Wendy the Warden will monitor all correspondence. There are two variations of the problem, with an active or a passive warden (see e.g. Fridrich *et al.* (2003a)). The passive warden model seems to be the one most commonly assumed in the literature. When Wendy is passive, she is only allowed to review, and not to modify, the messages. The passive warden is allowed to block communications, withdraw privileges, or otherwise punish Alice and Bob if she thinks they have abused their communication link (Figure 2.1). That could mean solitary confinement or the loss of communications. If Alice sends an encrypted message to Bob, it will be obviously suspicious, and Wendy, assuming the worst, will block it.

The seemingly less known scenario with an active warden is actually the original focus of Simmons (1983). The active warden can modify messages between Alice and Bob, aiming either to disable the hidden message (a Denial of Service attack), or forge a message which Bob will think is a genuine message from Alice, something which could serve to lure Bob into a trap.

Let's elaborate on the security problem of Alice and Bob. The centre of any security problem is some asset which needs protection; it can be any item of value, tangible or intangible. The main asset of Alice and Bob in the prisoners' problem is the privileges they are granted only as long as they are well-behaved and either do not abuse them or keep the abuse secret from Wendy. The most obvious privilege is the privilege to communicate. With a passive warden, the only threat is detection of

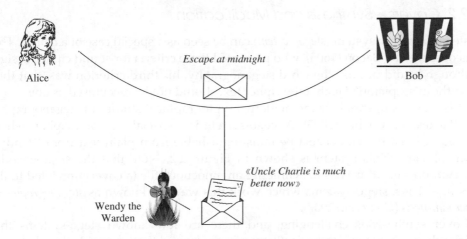

Figure 2.1 The prisoners' problem with passive warden

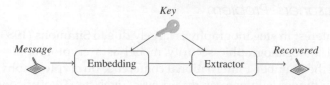

Figure 2.2   A typical (secret-key) system for steganography by cover synthesis

their abuse, leading to loss of privileges. Thus the passive warden scenario is a pure confidentiality problem, where some fact must be kept secret.

In the active warden scenario, there are three threats:

confidentiality detection of the abuse (as before);

availability loss of (illegal) messages due to modification; and

integrity misinformation (via messages forged by the Warden).

In the original work, Simmons (1983) addressed the integrity problem and suggested solutions for message authentication. The objective was to allow Bob to distinguish between genuine messages from Alice and messages forged by Wendy.

It should be noted that, in the passive warden model, the contents of the message is not a significant asset *per se*. Once Wendy suspects that an escape plan is being communicated, she will put Alice and/or Bob in solitary confinement, or take some other measure which makes it impossible to implement any plans they have exchanged. Thus, once the existence of the message is known, the contents have no further value, and keeping the very existence of the secret secret is paramount. Once that is lost, everything is lost.

2.2.2 Covers – Synthesis and Modification

A steganographic system (*stego-system*) can be seen as a special case of a cipher. This was observed by Bacon (1605), who presented three criteria for sound cipher design. Although he did not use the word steganography, his third criterion was 'that they be without suspicion', i.e. that the cipher text should not be recognised as one.

This view reappears in some modern, cryptographical studies of steganography, e.g. Backes and Cachin (2005). A stego-system is a special case of a cipher, where we require that the cipher text be indistinguishable from plain text from Wendy's point of view. This system is shown in Figure 2.2. Note that the stego-encoder (encryption algorithm) has to synthesise an innocuous file (a cover) unrelated to the message. Thus, stego-systems which work this way are known as *steganography by cover synthesis* (Cox *et al.*, 2007).

Cover synthesis is challenging, and there are few known stego-systems that efficiently apply cover synthesis. *Steganography by modification* avoids this challenge by taking an innocuous file, called the *cover*, as input to the stego-encoder in addition to the key and the message, as shown in Figure 2.3. This model is closely related to

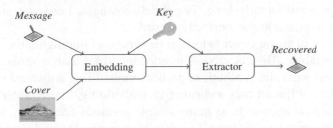

Figure 2.3 A typical (secret-key) system for steganography by modification

traditional communication systems, where the cover takes the role of a carrier wave. The message is then modulated by imperceptible modifications of the cover. The result is a steganogram which closely resembles the original cover. Many authors restrict the definition of steganography to cover only steganography by modification. For example, Cox *et al.* (2007, p. 425) 'define steganography as the practice of undetectably altering a Work to embed a message'.

In theory, cover synthesis is the most flexible and powerful approach, as the synthetic cover can be adapted to the message without restriction. In practice, however, media synthesis is so difficult to automate that almost all known practical systems use modification. There are notable exceptions though, such as the Spam-Mimic system[1] which imitates spam email messages. The author is not aware of any stego-system which synthesises images.

A third alternative is known as cover selection (Fridrich, 2009), which corresponds to using a code book in conventional cryptography, usually with many covers corresponding to each possible message. Each possible cover or cipher text is associated with a specific plain text, and the encoder will simply look up a cover to match the message. Cover selection is not always practical, as the code book required may be huge, but examples do exist.

Example 2.2.1 *Cover selection could be implemented by using a collection of thumbnail images as the cover. Each thumbnail is associated with an 8-bit message which is obtained by hashing the image using for instance an HMAC function (hash-based message authentication code) which can be truncated to the desired 8 bits. The encoder will split the message into 8-bit blocks. For each block it searches through a thumbnail bank, hashing each image until the correct hash value is found. Finally, all the thumbnails are collected, in the correct order, and transmitted as the steganogram. The decoder only has to hash each thumbnail to obtain the message.*

Properly done, cover selection steganography is impossible to detect. The steganogram is not only indistinguishable from an unmodified cover; it *is* an unmodified cover. In the thumbnail example above one could argue that the steganalyst can make inferences from the order or combination of images, especially

---

[1] http://www.spammimic.com/

if the message is sufficiently long. For short messages, however, steganalysis by cover selection is possible and perfectly secure.

The possibility of using cover selection or synthesis highlights the role the cover has in steganography. The objective is to encode, rather than to embed, and we can choose any cover we want, synthetic or natural. Embedding is just one way to encode. The current state of the art may indicate that embedding, that is cover modification, is the most practical approach, as many simple methods are known, but there is no other reason to prefer modification. Even steganography by modification allows one to be selective about the cover. It is not necessary that a stego-system by modification be secure or even work for arbitrary covers, as long as Alice determines in advance which covers are good.

2.2.3 Keys and Kerckhoffs' Principle

Early stego-systems depended entirely on the secrecy of the algorithm. The stego-encoder would take only the message and possibly a cover as input, and no secret key would be used. This is known as a *pure stego-system*. The problem with these systems is that once they are known, they become useless. It is straightforward for Wendy to try all known pure stego-decoders. Alice can only secretly send a message to Bob if Bob has a secret that Wendy does not know. If Wendy knows all that Bob knows she can decode the steganogram as he would.

The stego-systems shown in Figures 2.2 and 2.3 are *secret-key stego-systems*. The stego-encoder and -decoder are parameterised by a secret, known as the *key* shared by Alice and Bob. This key is the same concept that we know from encryption: it is any secret data which Bob has and Wendy has not, and which enables Bob to decode the message.

In cryptography one would always assume that the security depends only on the key, and that knowledge of the algorithm would not help Wendy to distinguish between a steganogram and an innocent message. This idea was very controversial when it was introduced by Kerckhoffs (1883), but is now universally accepted in the cryptography community and has been labelled *Kerckhoffs' Principle*. The steganography community is still debating whether and how Kerckhoffs' Principle applies to steganography.

Systems which adhere to Kerckhoffs' Principle can easily be instantiated with different secret keys and used for different applications. If one instance is compromised completely, this will not affect the other instances using different keys. This is a great advantage, because the key can easily be chosen at random, whereas algorithm design is a long and costly process. Without Kerckhoffs' Principle it would be impossible to get off-the-shelf cryptographic software, as the enemy would be able to buy the same software.

It should be noted that Kerckhoffs did not propose that all cryptographic algorithms ought to be made publicly available, and many experts in cryptography do indeed argue that, in practice, it may be wise to keep algorithms secret. He only said that encryption algorithms should be designed so that they could fall into enemy hands without inconvenience.

In reality, it has proved very hard to design stego-systems which are secure under Kerckhoffs' Principle. Excluding steganograms with extremely low payload, all known stego-systems seem to have been broken. Recent research on the *square-root law* of steganography (Ker *et al.*, 2008) indicates that any stego-system will be easy to break if it is used long enough, as the total length of the steganograms has to increase proportionally to the square of the total message lengths. This is very different from encryption, where the same cipher can be used indefinitely, with varying keys. Quite possibly, steganography will have to look beyond Kerckhoffs' Principle, but the debate is not over.

In practice, it will of course often be the case that Wendy has no information about Bob and Alice's stego-system. Hence, if no Kerckhoffs-compliant stego-system can be found, it is quite possible that Alice and Bob can get away with one which is not. Surely the feasibility of Kerckhoffs' Principle in encryption must have contributed to its popularity.

Public-key stego-systems have also been proposed (see e.g. Backes and Cachin (2005)), following the principles of public-key cryptography. Instead of Alice and Bob sharing a secret key, Bob generates a key pair (k_s, k_p), where the secret k_s is used as input to the decoder, and the public k_p is used as input to the encoder. Anybody knowing k_p can then send steganograms which only Bob can decode.

As far as we know, there are no practical public-key stego-systems which can handle more than very short messages. Common examples of public-key stego-systems require either extremely long covers or a perfect probability model of possible cover sources. Hopper (2004), for instance, suggests a public-key stego-system based on cover selection, where each message bit b is encoded as one so-called document D, which is simply selected from the cover source so that the decoding function matches the message bit, i.e. $f(D) = m$. Thus the steganogram becomes a sequence of documents, and although nothing is said about the size of each document, they must clearly be of such a length or complexity that the steganalyst cannot detect statistical anomalies in the distribution of document sequences.

2.2.4 LSB Embedding

One of the classic image steganography techniques is *LSB embedding*. It was initially applied to pixmap images, where each pixel is an integer indicating the colour intensity (whether grey scale or RGB). A small change in colour intensities is unlikely to be perceptible, and the stego-system will discard the least significant bit (LSB) of each pixel, replacing it with a message bit. The receiver can extract the message by taking the image modulo two.

There are a number of variations of LSB embedding. The basic form embeds sequentially in pixels or coefficients starting in the upper left corner, and proceeding in whichever direction is most natural for the compiler or interpreter. In Matlab and Fortran that means column by column, while in Python it is row by row by default.

This approach means that the hidden message will always be found in the same locations. Steganalysis becomes harder if the message is embedded in random pixels.

Embedding in random locations can be implemented by permuting a sequence of pixels using a pseudo-random number generator (PRNG) before embedding and inverting the permutation afterwards. The seed for the PRNG makes up the key which must be shared by Alice and Bob ahead of time, so that Bob can obtain the same permutation and know where to extract the message.

Example Python code for LSB embedding and extraction can be seen in Code Example 2.1. We shall give a brief introduction to Python in the next chapter. If key is given, it is used to seed a PRNG which generates a random permutation that is applied to the sequence of pixels before embedding. Thus the message bits are embedded in random locations. The embedding itself is done simply by subtracting the LSB (image modulo two) and then adding the message in its place.

Code Example 2.1 Python functions to embed and extract messages using LSB. The cover image X should be a 2-D numpy array of integers, and msg a 1-D numpy array of 0 and 1. The auxiliary routines im2sig and sig2im convert the 2-D image to a 1-D signal and vice versa, using the secret key to permute the elements

---

```
def embed(X,msg,key=None):
    (I,P,Sh) = im2sig( cover, key )
    L = len(msg)
    I[:L] −= I[:L]%2
    I[:L] += msg
    return sig2im(I,P,Sh)
def extract(X,key=None):
    (I,P,Sh) = im2sig( cover, key )
    return I % 2
def im2sig(cover,key=None):
    Sh = cover.shape
    if key != None:
      rnd.seed(key)
      P = numpy.random.permutation(S)
      I = np.zeros(S,dtype=cover.dtype)
      I[P] = cover.flatten()
      return (I,P,Sh)
    else:
      return (cover.flatten(),None,Sh)
def sig2im(I,P,Sh):
    if P == None: return I.reshape(Sh)
    else: return I[P].reshape(Sh)
```

Although LSB embedding most often refers to embedding in the spatial domain (pixmap), it can apply to any integer signal where a ± 1 change is imperceptible. In particular, it applies to JPEG images, but is then called JSteg in the key less version and Outguess 0.1 in the keyed version. The only special care JSteg takes is to ignore coefficients equal to 0 or 1; because of the great perceptual impact changes to or from 0 would cause. We will discuss JPEG steganography in further detail in Chapter 8.

Another variation of LSB, known as LSB matching, ± 1 embedding or LSB$\pm$, aims to randomise the distortion for each sample. The extraction function is the same for LSB and LSB$\pm$; taking the pixels modulo two returns the message. The embedding is different. When the pixel LSB matches the message bit to be embedded, the pixel is left unchanged as in LSB replacement. When a change is required, the embedding will add ± 1 with uniform probability to the pixel value, where LSB replacement would always add $+1$ for even pixels and -1 for odd ones.

The principles of LSB are used in a wide range of stego-systems in different variations. Few alternatives exist. More success has been achieved by applying LSB to different transforms of the image or by selectively choosing coefficients for embedding than by seeking alternatives to LSB. For instance, the recent HUGO algorithm of Pevný et al. (2010a) applies LSB$\pm$ to pixels where the change is difficult to detect.

2.2.5 Steganography and Watermarking

Steganography by modification, together with digital watermarking, form the area of information hiding, which in general aims to hide one piece of data (message) inside another piece of data (cover).

The difference between steganography and watermarking is subtle, but very important. Recall that Cox et al. (2007) defined steganography as *undetectably* altering a cover. Their definition of watermarking is

> the practice of imperceptibly altering a Work to embed a message that about Work.

Two points are worth noting. Firstly, watermarking is used to embed a message about the cover work itself. In steganography the cover is immaterial, and can in fact be created to suit the message, as it is in steganography by cover synthesis. Watermarking refers to a number of different security models, but what they have in common is an adversary seeking to break the connection between the Work and the embedded message. This could mean redistributing copies of the Work without the watermarked message, or forging a new Work with the same watermark.

We should also note the subtle difference between 'undetectably' and 'impercepti-bly'. Imperceptibility here means that the watermarked Work should be perceptually indistinguishable from the original Work for a human observer, something which is important when the cover is the asset. With the cover being of no relevance in steganography, there is also no need for imperceptibility in this sense. What we require is that Wendy will not be able to determine whether or not there is a secret message, i.e. undetectability. In watermarking the existence of the message is not necessarily a secret. In some application scenarios, we may even allow Wendy to extract it, as long as she cannot remove it.

Watermarking has been proposed for a number of different applications. Radio commercials have been watermarked for the purpose of broadcast monitoring, i.e. an automated system to check that the commercial is actually broadcast. There have been cases where broadcasters have cheated, and charged advertisers for airings which have not taken place, and broadcast monitoring aims to detect this. Clearly, in this application, the commercial and the watermark must be linked in such a way that the broadcaster cannot air the watermark without the commercial.

Another application is image authentication and self-restoration, which aims to detect, and possibly correct, tampering with the image. This has obvious applications in forensic photography, where it is important to ensure that an image is not modified after being taken. The watermark in authentication systems contains additional information about the image, so that if some sections are modified, the inconsistency is detected. In self-restoration, modified sections can be restored by using information from the watermarks of other sections.

Robust watermarking is used for copyright protection, where the adversary may try to distribute pirate copies of the Work without the watermark. Watermarking is said to be robust if it is impossible or impractical for the adversary to remove it. This is similar to the objective in the active warden scenario, where Wendy is not so much concerned with detecting the message as disabling it or falsifying it. Still, the main distinction between watermarking and steganography applies. In watermarking, the cover is the asset and the message is just a means to protect it. In steganography it is the other way around, and the active warden does not change that. Both cover synthesis and cover selection can be used in active warden steganography.

The difference between watermarking and steganography (by modification) may look minor when one considers the operation of actual systems. In fact, abuse of terminology is widespread in the literature, and some caution is required on the reader's part to avoid confusion. Part of the confusion stems from a number of works assessing the usability of known watermarking systems for steganography. That is a good question to ask, but the conclusion so far has been that watermarking is easy to detect.

In this book, we will strictly follow the terminology and definitions which seem to have been agreed by the most recent literature and in particular by the textbooks of Cox *et al.* (2007) and Fridrich (2009). The distinctions may be subtle, but should nevertheless be clear by comparing the security criteria and the threats and assets of the respective security models. In steganography, the message is the asset, while the cover is the asset in watermarking. Although many different threats are considered in watermarking, they are all related to the integrity confidentiality, integrity, or availability of the cover.

2.2.6 Different Media Types

Steganography, like watermarking, applies to any type of media. The steganogram must be an innocuous document, but this could be text, images, audio files, video files, or anything else that one user might want to send to another.

The vast majority of research in steganography focuses on image steganography, and the imbalance is even greater when we consider steganalysis. The imbalance can also be seen in watermarking, but not to the same extent. There are a number of noteworthy watermarking systems for speech signals (e.g. Hofbauer *et al.* (2009)) and video files. Part of the reason for this is no doubt that audio and video processing is somewhat more difficult to learn. Research in robust watermarking has also demonstrated that the strong statistical correlation between subsequent frames facilitates attacks which are not possible on still images, and this correlation is likely to be useful also in steganalysis.

There has been work on steganography in text documents, but this stems from other disciplines, like computational linguistics and information theory, rather than the signal- and multimedia-processing community which dominates multimedia steganalysis. Wayner (2002) gives several examples.

2.3 Steganalysis

Steganalysis refers to any attack mounted by a passive warden in a steganography scenario. It refers to Wendy's analysing the steganogram to obtain various kinds of information about Alice and Bob's correspondence. An active warden would do more than mere analysis and this is outside the scope of this book.

2.3.1 The Objective of Steganalysis

In the prisoners' problem with a passive warden, Wendy's objective is to detect the presence of secret communications. Any evidence that unauthorised communications are taking place is sufficient to apply penalties and would thus cause damage to Alice and Bob. Any additional information that Wendy may obtain is outside the model. The solution to her problem is a basic steganalyser, defined as follows.

Definition 2.3.1 *A basic steganalyser is an algorithm which takes a media file as input, and outputs either 'steganogram' or 'innocent'.*

Some authors argue that the basic steganalyser is only a partial solution to the problem, and that steganalysis should aim to design *message-recovery attacks* which are able to output also the secret message. These attacks clearly go beyond the security model of the prisoners' problem. There may very well be situations where Alice and Bob have additional assets which are not damaged by Wendy's learning of the existence of the secret, but would be damaged if the contents became known to the adversary.

There is, however, a second reason to keep message-recovery out of the scope of steganalysis. If Alice and Bob want to protect the contents of the communications even if Wendy is able to break the stego-system, they have a very simple and highly trusted solution at hand. They can encrypt the message before it is embedded, using a standard cryptographic tool, making sure that a successful message-recover attack on the stego-system would only reveal a cipher text. If the encryption is correctly applied, it is computationally infeasible to recover the plain text based on the cipher text. There is no significant disadvantage in encrypting the message before stego-encoding, as the computational cost of encryption is negligible compared to the stego-system. Thus, established cryptographic tools leave no reason to use steganography to protect the confidentiality (or integrity) of the contents of communications, and thus little hope of steganalysis to reveal it.

Of course, this does not mean that message-recovery is not an interesting problem, at least academically, but very little research has yet been reported, so the time for inclusion in a textbook has not yet come. There has been more work on other types of *extended steganalysis*, aiming to get different kinds of additional information about the contents of the steganogram.

Quantitative steganalysis aims to estimate the length of the secret message. *Key-recovery* attacks are obviously useful, as knowledge of the key would allow detection and decoding of any subsequent communication using the same key. Extended steganalysis also covers techniques to determine other information about the secret message or the technology used, such as the stego-algorithm which was used.

2.3.2 Blind and Targeted Steganalysis

Most early steganalysis algorithms were based on an analysis of individual known stego-systems, such as LSB in the spatial domain and JSteg. Such algorithms are known as *targeted* in that they target particular stego-systems and are not expected to work against any other stego-systems. Well-known examples include the pairs-of-values (or χ^2) test (Westfeld and Pfitzmann, 2000), sample pairs analysis (Dumitrescu *et al.*, 2002; Fridrich *et al.*, 2001b) and RS steganalysis (Fridrich *et al.*, 2001b). Targeted steganalysers can often be made extremely accurate against the target stego-system.

The opposite of targeted steganalysis is usually called *blind steganalysis*. We define blind steganalysis as those methods which can detect steganograms created by arbitrary and unknown stego-systems. Blind steganalysis can be achieved with one-class classifiers (cf. Section 11.5.1).

Most steganalysers based on machine learning are neither truly blind nor targeted. We will call these algorithms *universal*, because the classification algorithm applies universally to many, if not every, stego-system. However, with the exception of one-class classifiers, the classifiers have to be trained on real data including real steganograms. Thus the training process is targeted, while the underlying algorithm is in a sense blind. Some authors consider universal steganalysis to be blind, but we consider this to be misleading when Wendy has to commit to particular stego-systems during training.

Each of blind, universal and targeted steganalysers have important uses. A firm believer in Kerckhoffs' Principle may argue that targeted steganalysis is the right approach, and successful design of a targeted steganalyser renders the target stego-system useless. Indeed, targeted steganalysis is a very useful tool in the security analysis of proposed stego-systems. In practice, however, Wendy rarely knows which algorithm Alice has used, and it may be computationally infeasible for Wendy to run a targeted steganalyser for every conceivable stego-system. It would certainly make her task easier if we can design effective blind steganalysers.

The concept of targeted blind steganalysis has been used by some authors (e.g. Pevný *et al.*, 2009b). It refers to universal steganalysers whose training has been targeted for specific use cases. This need not be limited to a particular stego-system, but could also assume a particular cover source.

Universal steganalysers are limited by the data available for training. If the training data are sufficiently comprehensive, they may, in theory, approach blind steganalysis. However, computational complexity is a serious issue as the training set grows. A serious limitation of steganalysers based on machine learning is that they depend on a range of characteristics of the training set. Taking images as an example, these characteristics include JPEG compression factors, lighting conditions, resolution, and even characteristics of the source camera (sensor noise). The more information Wendy has about possible cover images, the more accurately she can train the steganalyser.

It is useful to distinguish between a steganalytic algorithm or steganalyser, which is a fundamental and atomic technique for steganalysis, and a steganalytic system, which is a practical implementation of *one or more* algorithms. A common approach to achieve a degree of blindness is to combine multiple steganalytic algorithms to cover a range of stego-algorithms in a complex system. As the individual algorithms are simpler than the system, they can be evaluated more thoroughly, and mathematical theory can be designed. We will return to the implementation issues and complex steganalytic systems in Chapter 14.

2.3.3 Main Approaches to Steganalysis

Steganalysis ranges from manual scrutiny by experts to automated data analysis. We can distinguish between four known classes of image steganalysis. Ordered by the need for manual action, they are

1. visual steganalysis,
2. structural steganalysis,
3. statistical steganalysis and
4 learning steganalysis.

Visual and structural steganalysis are well suited for manual inspection, and visual steganalysis completely depends on a subjective interpretation of visual data. Statistical and learning steganography, on the other hand, are well suited for automated calculation, although statistical steganalysis will usually have an analytic basis.

Visual Steganalysis

Visual steganalysis is the most manual approach, as Wendy would use her own eyes to look for visual artifacts in the image itself. In addition to inspecting the full uncompressed image, she can also examine transforms of the image. One of the most popular tools for visual inspection is the LSB plane plotted as a two-tone image, as illustrated

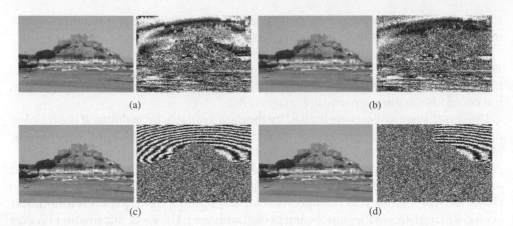

(a)   (b)

(c)   (d)

Figure 2.4 Clean images and steganograms with corresponding LSB bit plane: (a) clean GIF image; (b) EzStego steganogram; (c) clean grey-scale image; (d) LSB steganogram. The steganograms have 6772 bytes embedded out of a capacity of 13 312 bytes (photograph of Mont Orgueil, Jersey)

in Figure 2.4. The source image was captured by a Nikon D80 and stored with the highest resolution and highest compression quality, albeit not in raw format. The image was downsampled by a factor of 10 and converted using ImageMagick convert. Both the 8-bit grey-scale image and the GIF image were obtained directly from the source image. The messages embedded are unbiased, random bit strings. EzStego embeds in random locations. The embedding in the grey-scale image is in consecutive images, and we can see how it clearly covers the left-hand half of the image.

It should be noted that the grey-scale example in Figure 2.4 is not necessarily typical. Most images we have inspected do not display as much visual structure in the LSB plane as this one. However, this structure is described as typical by other authors focusing on palette images, including Westfeld and Pfitzmann (2000) and Wayner (2002).

Visual steganalysis is obviously very flexible, as it can pick up on new and previously unknown artifacts. However, not all cover images exhibit the same visual structure, making the approach very cover-dependent.

Structural Steganalysis

Structural steganalysis looks for give-away signs in the media representation or file format. A classic and blatant example is some of the software packages of the 1990s, which inserted the name of the stego software in the comment field of JFIF (JPEG) files. The variations in how the standard is implemented may also give room for forensic analysis, tracing a given file back to particular software packages including stego-embedders. The variations may include both bugs and legitimate design choices.

Vulnerabilities to structural attacks include artificial constraints, like only allowing images of 320 × 480 pixels as Hide and Seek version 4.1 did (Wayner, 2002).

Although this does not in itself prove anything, it gives a valuable clue as to what other artifacts may be worth investigating. The colour palette, for instance in GIF images, is a source of many other vulnerabilities. Many steganography systems reorder the palette to facilitate simple, imperceptible embedding. S-Tools is an extreme example of this, as it first creates a small, but optimal 32-colour palette, and then adds, for each colour, another seven near-identical colours differing in only one bit (Wayner, 2002). Any palette image where the colours come in such clusters of eight is quite likely to be a steganogram.

Another structural attack is the JPEG compatibility attack of Fridrich *et al.* (2001a). This attack targets embedding in pixmap images where the cover image has previously been JPEG compressed. Because JPEG is lossy compression, not every possible pixmap array will be a possible output of a given JPEG decompression implementation. The attack quite simply checks if the image is a possible decompression of JPEG, and if it isn't it has to be due to some kind of post-compression distortion which is assumed to be steganographic embedding.

Structural steganalysis is important because it demonstrates how vulnerable quick and dirty steganography software can be. In contrast, a steganographer can easily prevent these attacks by designing the software to avoid any artificial constraints and mimic a chosen common-place piece of software. It is possible to automate the detection of known structural artifacts, but such attacks become very targeted. An experienced analyst may be able to pick up a wider range of artifacts through manual scrutiny.

Statistical Steganalysis

Statistical steganalysis uses methods from statistics to detect steganography, and they require a statistical model, describing the probability distribution of steganograms and/or covers theoretically. The model is usually developed analytically, at least in part, requiring some manual analysis of relevant covers and stego-systems. Once the model is defined, all the necessary calculations to analyse an actual image are determined, and the test can be automated.

A basic steganalyser would typically use statistical hypothesis testing, to determine if a suspected file is a plausible cover or a plausible steganogram. A quantitative steganalyser could use parameter estimation to estimate the length of the embedded message. Both hypothesis testing and parameter estimation are standard techniques in statistics, and the challenge is the development of good statistical models.

Statistical modelling of covers has proved to be very difficult, but it is often possible to make good statistical models of steganograms from a given stego-system, and this allows targeted steganalysers.

Learning Steganalysis

In practice it is very often difficult to obtain a statistical model analytically. Being the most computerised class of methods, learning steganalysis overcomes the analytic challenge by obtaining an empirical model through brute-force data analysis.

The solutions come from areas known as pattern recognition, machine learning, or artificial intelligence.

Most of the recent research on steganalysis has been based on machine learning, and this is the topic of the present book. The machine is presented with a large amount of sample data called a training set, and is started to estimate a model which can later be used for classification. There is a grey area between statistical and learning techniques, as methods can combine partial analytic models with computer-estimated models of raw data. Defining our scope as learning steganalysis therefore does not exclude statistical methods. It just means that we are interested in methods where the data are too complex or too numerous to allow a transparent, analytic model. Machine learning may come in lieu of or in addition to statistics.

2.3.4 Example: Pairs of Values

A pioneer in statistical steganalysis is the pairs-of-values (PoV) attack which was introduced by Westfeld and Pfitzmann (2000). It may be more commonly known as the χ^2 attack, but this is somewhat ambiguous as it refers to the general statistical technique for hypothesis testing which is used in the attack. The attack was designed to detect LSB embedding, based on the image histogram.

To understand pairs-of-values, it is instructive to look at the histograms of a cover image and a steganogram where the cover image has been embedded with LSB embedding at full capacity, i.e. one bit per pixel. An example, taken from 8-bit grey-scale images, is shown in Figure 2.5. The histogram counts the number of pixels for each grey-scale value. Note that the histogram of the steganogram pre-dominantly has pairs of bars at roughly equal height.

If we look at what happens in LSB embedding, we can explain these pairings. A pixel with value $2i$ in the cover image can either remain at $2i$ or change to $2i + 1$. A pixel of value $2i + 1$ can become either $2i$ or $2i + 1$. Thus $(2i, 2i + 1)$ forms a pair of values for each $i = 0, 1, \ldots, 127$. LSB embedding can change pixel values within a pair, but cannot change them to values from a different pair. The embedded message seems to have an even distribution of 0 and 1, which is to be expected in a random or encrypted message, and this results in an approximately even distribution of $2i$ and $2i + 1$ for each i in the steganogram.

We can use statistical hypothesis testing to check if a suspected image is a plausible steganogram, by checking if the frequencies of $2i$ and $2i + 1$ are plausible under an even probability distribution. The pairs-of-values attack uses the χ^2 test to do this. Let h_m be the number of pixels of intensity m. In a steganogram (at full capacity) we expect an even distribution between the two values in a pair, thus $E(h_{2l}) = (h_{2l} + h_{2l+1})/2$. Following the standard framework for a χ^2 test (see e.g. Bhattacharyya and Johnson (1977)), we get the statistic

$$S_{\mathrm{PoV}} = \sum_{l=0}^{127} \frac{[h_{2l} - \frac{1}{2}(h_{2l} + h_{2l+1})]^2}{\frac{1}{2}(h_{2l} + h_{2l+1})}. \tag{2.1}$$

Consulting the χ^2 distribution, we can calculate the corresponding p-value.

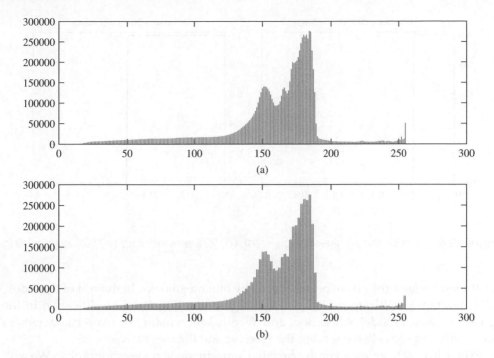

Figure 2.5 Histograms of grey-scale images: (a) clean image; (b) steganogram

The pairs-of-values test was first designed for consecutive LSB embedding. Then we can perform the test considering only the first N pixels of the image. In Figure 2.6, we have plotted the p-value for different values of N. The p-value is the probability of seeing the observed or a larger value of S_{PoV} assuming that the image is a steganogram. We can easily see that when we only consider the pixels used for embedding, we get a large p-value indicating a steganogram. When we start considering more pixels than those used for embedding, the p-value drops quickly. Thus we can use the plot to estimate the message length.

The basic form of the χ^2 attack is only effective against LSB replacement in consecutive pixels. However, the idea has been extended several times to attack other stego-systems and their variations. Very early on, Provos and Honeyman (2001) introduced the extended χ^2 attack to attack embedding in random locations. Later, Lee *et al.* (2006, 2007) created the category attack which is very effective against embedding in the JPEG domain.

2.4 Summary and Notes

We have summarised the wide variety of techniques and approaches in steganography and steganalysis, and it is time to define our scope. Some topics are excluded because they do not currently appear to be sufficiently mature, others have been

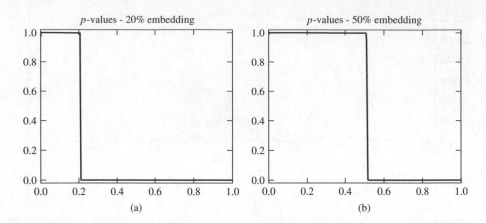

Figure 2.6   Plots of the p-values for a χ^2 test: (a) 20% embedding; (b) 50% embedding

excluded because they have been covered by other textbooks. In the rest of this book, steganography will refer to image steganography using cover modification in the passive warden model. Alice, Bob, and Wendy will remain our main characters, to personify respectively the sender, the receiver and the steganalyst.

Fridrich (2009) gives a comprehensive introduction to steganography. We will focus on steganalysis using machine learning in this steganography model only. Some statistical and visual steganalysis methods will be presented as examples, and a somewhat wider overview can be found in Fridrich (2009). For the most part we will limit the discussion to basic steganalysis. When we investigate extended steganalysis later in the book, we will use appropriately qualified terms.

3

Getting Started with a Classifier

In Part II of the book, we will survey a wide range of feature vectors proposed for steganalysis. An in-depth study of machine learning techniques and classifiers will come in Part III. However, before we enter Part II, we need a simple framework to test the feature vectors. Therefore, this chapter starts with a quick overview of the classification problem and continues with a hands-on tutorial.

3.1 Classification

A *classifier* is any function or algorithm mapping objects to classes. Objects are drawn from a population which is divided into disjoint classes, identified by class *labels*. For example, the objects can be images, and the population of all possible images is divided into a class of steganograms and a class of clean images. The steganalytic classifier, or *steganalyser*, could then take images as input, and output either 'stego' or 'clean'.

A class represents some property of interest in the object. Every object is a member of one (and only one) class, which we call its *true class*, and it is represented by the true label. An ideal classifier would return the true class for any object input. However, it is not always possible to determine the true class by observing the object, and most of the time we have to settle for something less than ideal. The classifier output is often called the *predicted class* (or predicted label) of the object. If the predicted class matches the true class, we have correct classification. Otherwise, we have a classification error.

Very often, one class represents some specific property and we aim to detect the absence or presence of this property. The presence is then considered as 'positive'

Machine Learning in Image Steganalysis, First Edition. Hans Georg Schaathun.
© 2012 John Wiley & Sons, Ltd. Published 2012 by John Wiley & Sons, Ltd.

and absence as 'negative'. It is therefore natural to speak of two different error types. A positive prediction for a negative object is called a *false positive*, and conversely, a negative prediction for a positive object is a *false negative*. Very often, one error type is more serious than the other, and we may need to assess the two error probabilities separately.

Steganalysis aims to detect the presence of a hidden message, and thus steganograms are considered positive. Falsely accusing the correspondents, Alice and Bob, of using steganography is called a false positive. Failing to detect Alice and Bob's secret communication is called a false negative. Accusing Alice and Bob when they are innocent (false positive) is typically considered to be a more serious error than a false negative.

3.1.1 Learning Classifiers

A typical classifier system is depicted in Figure 3.1. It consists of a training phase, where the machine learns, and a test phase, where the learning is practised. In the centre is the *model*, which represents the knowledge acquired during training. The model is output in the training phase and input in the test phase.

It is rarely feasible for the classifier to consider the object as a whole. Instead, a *feature vector* is extracted to represent the object in the classifier. Each feature is (usually) a floating point value, and it can be any kind of measure taken from the object. Selecting a good feature vector is the main challenge when machine learning is applied to real problems. Obviously, the same method of feature extraction must be used in training and in testing.

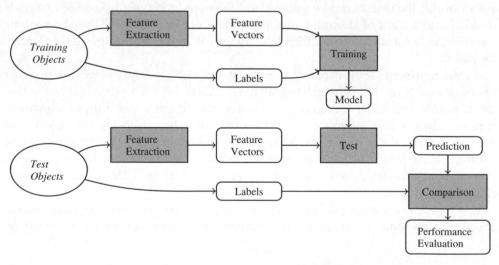

Figure 3.1 A learning classifier system. The grey boxes represent operations or algorithms, while the white boxes represent data

Before the classifier system can actually classify objects, it has to be trained on some training set. The figure depicts *supervised learning*, which means that the training algorithm receives both the feature vector and the true class label for each training object.

Testing is somewhat ambiguous. In practical use, we want to use the system to test objects in order to determine their class. In this case we have an object with unknown class, and its feature vector is input to the system, together with the model. The classifier outputs a class *prediction*, which we assume is its class.

Before actual use, it is necessary to evaluate the classifier to estimate how accurately it will classify objects. This evaluation is also commonly known as the *test phase*. When we test the classifier, we need a set of test objects with known labels, but the labels are not made available to the classifier. These objects are tested as described above and the resulting predictions are compared with the known labels, and the number of errors are counted. This allows us to calculate error rates and other performance measures.

3.1.2 Accuracy

When we review feature vectors for steganalysis in Part II, we will use one simple performance heuristic, namely the *accuracy*, which is defined as the rate of correct classifications during testing, or equivalently

$$\hat{A} = 1 - \frac{F_P + F_N}{T_P + T_N}, \tag{3.1}$$

where F_P and F_N are the number of false positives and false negatives in the test respectively, and T_P and T_N are the total number of positives and negatives. Obviously, the error rate is $1 - \hat{A}$.

Normally, we expect the accuracy to be in the range $0.5 \leq \hat{A} \leq 1$. A classifier which randomly outputs 'stego' or 'clean' with 50/50 probability, that is a *coin toss*, would get $\hat{A} = 0.5$ on average. If we have a classifier C with $\hat{A} \ll 0.5$, we can design another classifier C' which outputs 'stego' whenever C says 'clean' and vice versa. Then C' would get accuracy $\hat{A}' = 1 - \hat{A} \gg 0.5$. It is often said that a classifier with $\hat{A} \ll 0.5$ has good information but applies it incorrectly.

Accuracy is not the only performance measure we could use, nor is it necessarily the best one. Accuracy has the advantages of being simple to use and popular in the steganalysis literature. A problem with accuracy is that it depends on the class skew, that is the relative size of the classes. For example, consider a classifier which always outputs a negative. Clearly, this would not be a useful algorithm, and if there is no class skew, that is $T_P = T_N$, then $\hat{A} = 50\%$. However, if the positive class is only 1% of the population, i.e. $T_N = 99T_P$, then we would suddenly get 99% accuracy with the very same classifier.

The accuracy is worthless as a measure without knowledge of the class skew. When we quote accuracy in this book, we will assume that there is no class skew ($T_P = T_N$), unless we very explicitly say otherwise. We will explore other performance measures in Chapter 10.

3.2 Estimation and Confidence

The accuracy $\hat{A}$ as defined above is an empirical measure. It describes the outcome of an experiment, and not the properties of the classifier. Ideally, we are interested in the true accuracy A, which is defined as the probability of correct classification, but this, alas, would require us to test every possible image from a virtually infinite universe. Clearly impossible.

The empirical accuracy $\hat{A}$ is an estimator for the true accuracy A. Simply put, it should give us a good indication of the value of A. To make a useful assessment of a classifier, however, it is not sufficient to have an estimate of the accuracy. One also needs some information about the quality of the estimate. How likely is it that the estimate is far from the true value?

We will discuss this question in more depth in Chapter 10. For now, we shall establish some basic guidance to help us through Part II. An estimator $\hat{\theta}$ for some unknown parameter θ is a stochastic variable, with some probability distribution. It has a mean $\mu_{\hat{\theta}}$ and a variance $\sigma_{\hat{\theta}}^2$. If $\mu_{\hat{\theta}} = \theta$, we call $\hat{\theta}$ an unbiased estimator of θ. In most cases, it is preferable to use the unbiased estimator with the lowest possible variance.

The standard deviation $\sigma_{\hat{\theta}}$ is known as the *standard error* of the estimator, and it is used to assess the likely estimation error. We know from statistics that if $\hat{\theta}$ is normally distributed, a 95.4% error bound is given as $\pm 2\sigma_{\hat{\theta}}$. This means that, prior to running the experiment, we have a 95.4% chance of getting an estimation error

$$|\hat{\theta} - \theta| \leq 2\sigma_{\hat{\theta}}.$$

In practice, σ is unknown, and we will have to estimate the standard error as well.

The test phase is a series of Bernoulli trials. The test of one image is one trial which produces either correct classification, with probability A, or a classification error, with probability $1 - A$. The number of correct classifications in the test phase is binomially distributed, and we denote it as $C \sim B(A, N)$. Then $\hat{A} = C/N$, and it follows that the expected value is

$$E(\hat{A}) = \frac{E(C)}{N} = \frac{AN}{N} = A, \tag{3.2}$$

showing that $\hat{A}$ is an unbiased estimator for A. We know that the variance of $\hat{A}$ is $A(1 - A)/N$, and the customary estimator for the standard error is

$$\widehat{SE}(\hat{A}) = \sqrt{\frac{\hat{A}(1 - \hat{A})}{N}}. \tag{3.3}$$

By the normal approximation of the binomial distribution, we have approximately a 95.4% chance of getting an estimation error less than ϵ prior to running the experiment, where

$$\epsilon = 2\sqrt{\hat{A}(1 - \hat{A})/N}.$$

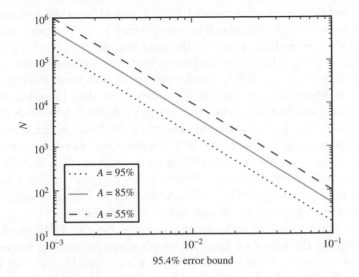

Figure 3.2 The number of test samples required to obtain a given 95.4% error margin

Note the dependency on the number of test images N. If we want to tighten the error bound, the obvious approach is to increase N. This dependency is illustrated in Figure 3.2.

It is useful to look at the plot to decide on a test set size which gives a satisfactory error bound. In Part II we have used a test set of 4000 images (2000 for each class). When estimating accuracies around 85% this gives an error bound of ±0.011. In other words, almost 5% of the time the estimate is off by more than 1.1 percentage points. This is important when comparing feature vectors. If the estimated accuracies don't differ by more than 1.1 percentage points or so, then we should consider them approximately equal. Or if we really want to show that one method is better than the other, we could rerun the test with a larger test set, assuming that a larger test set is available.

The error bound is somewhat smaller for higher accuracies, but this is deceptive. First note that the classification error probabilities $E_{class} = 1 - A$ is as important as the accuracy A. When studying classification error probabilities of a few percent, or equivalently, accuracies close to 100%, a difference of one percentage point is much more significant than it is for accuracies less than 90%. Focusing on ϵ, the relative error bound, that is

$$E_{rel} = \frac{\epsilon}{E_{class}},$$

is larger for small error probabilities.

For example, suppose you have a method with an estimated accuracy $\hat{A} = 99\%$ and an error bound of 0.011. Then introduce a new method, which we think has a better accuracy. How can you make a convincing case to argue that it is better?

If we were studying accuracies around 85%, it would be quite conceivable that a better method could improve the estimated accuracy by more than the error bound, but starting at 99% even 100% is within the error bound.

It is useful to design the experiments so that the accuracies are not too close to 1. In steganalysis, this probably means that we do not want to embed very long messages, as shorter messages are harder to detect. By thus lowering the accuracy, different methods and feature vectors become more distinguishable. A common rule of thumb, in some areas of simulation and error probability estimation, is to design the experiment in such a way that we get some fixed number r of error events (classification errors in our case); $r = 100$ errors is common. This is motivated by the fact that the standard error on the estimate of $\log E_{\text{class}}$ is roughly linear in r and independent of N (MacKay, 2003, p. 463). Applying this rule of thumb, it is obvious that an increase in A requires an increase in the test set size N.

There is a second reason to be wary of accuracies close to 1. We used the normal approximation for the binomial approximation when estimating the error bound, and this approximation is only valid when A is not too close to 1 or 0, and N is sufficiently large. When the 95.4% error bound does not bound A away from 1, or in other words when $\hat{A} + 2\widehat{\text{SE}}(\hat{A}) > 1$, it may well be a sign that we are moving too close to 1 for the given size of test set N. Such estimates of accuracy should be taken with a big pinch of salt.

3.3 Using libSVM

Support vector machines (SVM) were introduced by Boser *et al.* (1992) and Cortes and Vapnik (1995), and quickly became very popular as a classifier. Many implementations of SVM are available, and it is easy to get started as an SVM user. The libSVM package of Chang and Lin (2001) is a good choice, including both command line tools and APIs for most of the popular programming languages, such as C, Python and Matlab. For the initial experiments throughout Part II, we will follow the recommendations of Hsu *et al.* (2003).

3.3.1 Training and Testing

The first task is to compile training and test sets. For each object, we have a class label l, usually $l \in \{0, 1\}$, and a feature vector $(f_1, f_2, \ldots, f_n)$. Using libSVM on the command line, it takes plain text files as input. Each line represents one object and can look, for instance, like this

```
1 1:0.65 3:0.1 4:-9.2 10:2
```

which is interpreted as

$$l = 1, f_1 = 0.65, f_3 = 0.1, f_4 = -9.2, f_{10} = 2,$$

and $f_i = 0$ for any other i. The first number is the class label. The remaining, space-separated, elements contain the feature number, or coordinate index, before the colon and the feature itself after the colon. Zero features may be omitted, something which is useful if the data are sparse, with few non-zero features.

To get started, we must create one such file with training data and one containing test data. In the sequel, we assume they exist and are called `train.txt` and `test.txt`, respectively. In the simplest form, we can train and test libSVM using the following commands:

```
$ svm-train train.txt
*

[...]
$ svm-predict test.txt train.txt.model predict1.txt
Accuracy = 50.2959% (170/338) (classification)
```

This uses radial basis functions (RBF) as the default kernel. We will not discuss the choice of kernel until Chapter 11; the default kernel is usually a good choice. However, as we can see from the accuracy of $\approx 50\%$, some choice must be bad, and we need some tricks to improve this.

A potential problem with SVM, and many other machine learning algorithms, is that they are sensitive to scaling of the features. Features with a large range will tend to dominate features with a much smaller range. Therefore, it is advisable to scale the features so that they have the same range. There is a libSVM tool to do this, as follows:

```
$ svm-scale -l -1 -u +1 -s range train.txt > train-scale.txt
$ svm-scale -r range test.txt > test-scale.txt
$ svm-train train-scale.txt
*

[...]
$ svm-predict test-scale.txt train-scale.txt.model predict2.txt
Accuracy = 54.7337% (185/338) (classification)
```

The first command will calculate a linear scaling function $f' = af + b$ for each coordinate position, and use it to scale the features to the range specified by the -1 and -u options. The scaling functions are stored in the file `range` and the scaled features in `train-scale.txt`. The second command uses the parameters from the `range` file to scale the test set.

The last two commands are just as before, to train and test the classifier. In this case, scaling seems to improve the accuracy, but it is still far from satisfactory.

3.3.2 Grid Search and Cross-validation

Choosing the right parameters for SVM is non-trivial, and there is no simple theory to aid that choice. Hsu *et al.* (2003) recommend a so-called grid search to identify the SVM parameter C and kernel parameter γ. This is done by choosing a grid of trial values for (γ, C) in the 2-D space, and testing the classifier for each and every parameter choice on this grid. Because the expected accuracy is continuous in the parameters, the trial values with the best accuracy will be close to optimal.

The grid search is part of the training process, and thus only the training set is available. In order to test the classifier on the grid points, we need to use only a part of the training set for training, reserving some objects for testing. Unless the training set is generous, this may leave us short of objects and make the search erratic. A possible solution to this problem is cross-validation.

The idea of cross-validation is to run the training/testing cycle several times, each time with a different partitioning into a training set and a test set. Averaging the accuracies from all the tests will give a larger sample and hence a more precise estimate of the true accuracy. Each object is used exactly once for testing, and when it is used for testing it is excluded from training.

In n-fold cross-validation, we split the set into n equal-sized sets and run n repetitions. For each round we reserve one set for testing and use the remaining objects for training. The cross-validation estimate of accuracy is the average of the accuracies for the n tests.

The libSVM package includes a script to perform grid search. Both C and γ are searched on a logarithmic scale, and to save time, a coarse search grid is used first to identify interesting areas before a finer grid is applied on a smaller range. The grid search may seem overly banal and expensive, but it is simple and effective. Computational time may be saved by employing more advanced search techniques, but with only two parameters (C and γ) to tune, the grid search is usually affordable. The grid search can be visualised as the contour plot in Figure 3.3, which has been generated by the grid search script. Each contour represents one achieved accuracy in the (C, y) space.

The grid search script can be used as follows:

```
$ grid.py train-scale.txt
[local] 5 -7 82.4 (best c=32.0, g=0.0078125, rate=82.4)
[...]
[local] 13 -3 74.8 (best c=2048.0, g=0.001953125, rate=84.3)
2048.0 0.001953125 84.3
```

The last line of output gives the result. The optimal parameters are found to be $C = 2048$ and $\gamma = 0.001953125$, and the cross-validation accuracy is 84.3% . The default is 5-fold cross-validation, so 80% of the training set is used for cross-validation training. When training the classifier, we can specify the C and γ values that we found, as follows:

```
$ svm-train -c 2048 -g 0.001953125 train-scale.txt
..............................*........*
[...]
$ svm-predict test-scale.txt train-scale.txt.model predict3.txt
Accuracy = 85.503% (289/338) (classification)
```

As we can see, testing confirms the improved accuracy, and the accuracy estimated by cross-validation is fairly good. Throughout Part II we will use the above procedure to evaluate the various feature vectors.

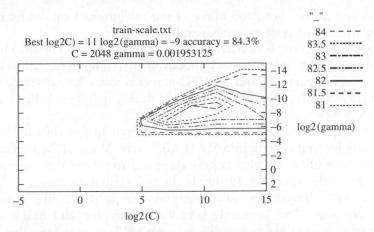

Figure 3.3 Contour plot of the cross-validation results from the grid search. The `grid.py` script was slightly modified to make the contour plot more legible in print

3.4 Using Python

Throughout the book, we will illustrate the techniques we cover by examples and demonstrations in Python. In this chapter we will give a very brief introduction to Python and the standard libraries that we require in the book, including the imaging library (PIL) and the libraries for scientific computing (pylab). This is intended to be just sufficient to comprehend the code examples used in the book.

3.4.1 Why we use Python

Python is a general-purpose, interpreted, object-oriented programming language. In addition to object-oriented features, it also supports functional features and syntax, making it extremely flexible. The interpreter can be run interactively, making it relatively easy to experiment and prototype. The interpreters also support a compiled byte-code format.

In the planning of this book, both MATLAB® and Python were considered. MATLAB®, with its very simple and intuitive syntax for interactive scientific computing, is very wide spread in the steganalysis community. Many authors have published Matlab code to demonstrate their methods. Unfortunately, Python's syntax for matrix algebra and scientific computing is not quite as intuitive and straight forward as Matlab's.

There are two important reasons why we chose to use Python. Firstly, contrary to MATLAB® it was designed as a general-purpose language from the very beginning. This generally makes it easier to write clean and structured code, the C interface is easier to understand, and running Python non-interactively in batch is more straight forward and robust. Secondly, there is the price; Python interpreters, including the extensive APIs for scientific computing, are available as free software and free of charge. Although MATLAB® itself is quite affordable if you run it only on a workstation and do not need too many of the toolboxes, it can easily blow your budget with extensive use of multiple, parallel instances.

Several of the experiments underlying this book only became feasible by running parallel, independent batch processes on about 50–100 computers. Python lends itself well to this sort of operation, and the CPU power is also a free resource in student labs which are otherwise idle both at night and during holidays. In this scenario our Matlab licences were insufficient.

Readers who prefer to use MATLAB® should not find it difficult to translate the code examples in this book into MATLAB® code. Many of the Python libraries for scientific computing are also largely designed to follow the user interface of MATLAB® as much as possible, further facilitating code translation.

We never really considered using compiled languages for the examples and companion software. This is mainly because prototyping and trial and error is more cumbersome and time-consuming in compiled languages, and that is a major drawback in a research project. We hope that our readers will play around with some of our examples in an interactive interpreter such as `ipython`.

3.4.2 Getting Started with Python

The examples in this book have been tested and run on a standard Unix/Linux system (Ubuntu), where all the necessary libraries are available in the package management system, making it easy to get started. All the software is open source, and hence it can be installed on other operating systems as well, such as Windows or Mac OS.

Most of the scientific computing libraries are implemented in C, and will only work with the standard C implementation of Python. The examples in the book will not work, for instance with the Java implementation known as `jython`. Python can be used both interactively and non-interactively, but for interactive use, it is worth installing a special interactive interpreter, `ipython`, which provides additional features such as command history, word completion, and inline help with paging and search functions.

Objects can be defined either in separate files, or modules. In Python, lists, classes and other types, and functions, are objects in the sense that they can be passed as arguments and stored in composite objects like arrays. To define a function, we use the `def` keywords:

```
def func(x,y):
    "A banal sample function to multiply two numbers."
    return x*y
```

Note that Python has no keywords to mark the beginning and end of a block; nor symbols to mark the end of a statement. Instead, a statement ends with a line break, and indentation marks the blocks. Everything which is indented more than the previous line forms a block. The string at the beginning of the definition is the *docstring*, which is intended for user documentation. At least in `ipython`, the documentation can be accessed conveniently by calling `help` on any object:

```
help func
```

Python has some support for object-orientation, and classes (types) are defined as follows. Methods are functions defined within the class.

```
class myclass(object):
    "A very simple example class, storing a name."
    def __init__(self,name):
        self.name = name
    def get_name(self):
        return self.name
obj = myclass("John Doe")
name = obj.get_name()
```

Calling a method from an object, the object itself is passed (implicitly) as the first argument, and this is the only way the method knows which object it belongs to. This argument is customarily called `self`.

Like most functional languages, Python has native support and simple syntax for list processing. Lists can be defined either by enumerating all the elements or by list comprehension as follows:

```
X = [ 0, 1, 4, 6, 10, 15, 20, 30, 40, 50 ]
L = [ float(x)/25 for x in X ]
```

Lists can also be used as iterators in loops, like

```
for i in L:
    print i
```

Objects may be defined directly on the command line, or in separate files, which form modules. Python files must have a name ending in `.py`. To load the definitions from a file, one can use

```
from file import *
```

Python is dynamically typed. That is, all the objects will have a type, and one can identify the type using the `isinstance()` or `type()` methods. However, there is no static type checking, so any variable can be assigned objects of any type. This may require some care in programming. Many errors which cause type errors in compilation or early in interpreting may go undetected for a long time in Python, possibly causing logical errors which are hard to trace. In contrast, it gives a lot of flexibility, and objects may be used interchangeably irrespective of their type if they implement the same interface as relevant in the context. Furthermore, interfaces do not have to be formally defined. If it quacks like a duck, it is assumed to be a duck. This is called 'duck typing' in the jargon.

3.4.3 Scientific Computing

There is a good supply of libraries for scientific computation in Python. Most importantly, `numpy` provides data types and basic operations for matrix algebra. The `matplotlib` library provides plotting functions, largely cloning the API of

MATLAB®. Finally, `scipy` provides additional functions for scientific computations, building on top of `numpy`. These three libraries are often presented together, as the `pylab` library or the pylab project. Also, `ipython` is considered part of the pylab project.

The most important feature of `numpy` is the `array` class. The libraries must be imported with the `import` statement, and it is customary to give them local shorthand names with the `as` clause, as follows:

```
import numpy as np
import scipy   as sp
A = np.array( [ [ 1.0, 1.5, 2.0 ], [ 0.5, 2.5, 2.0 ] ] )
B = np.random.randn( 2,3 )
```

In the example, A is a 2×3 array with the given entries and B becomes a 2×3 random array with standard normal distribution $\mathcal{N}(0, 1)$.

Arithmetic operations on arrays are element-wise, whereas matrix arithmetic is available as methods. For instance, the following print the element-wise product and the dot product, respectively:

```
print A * B
print np.dot( A, B.transpose() )
```

Functions to create various types of plots are available in the `matplotlib.pyplot` package, for instance using

```
import matplotlib.pyplot as plt
plt.plot( X, np.log(X), "k−x" )
plt.savefig( "testplot.pdf" )
```

Note that many functions, such as the logarithm `np.log` above, work element-wise on arrays. Naturally, there is a wide range of different types of plots, and it is worth browsing the `matplotlib` documentation (Hunter *et al.*, 2011).

3.4.4 Python Imaging Library

There are two libraries that can handle images in Python. Firstly, `matplotlib` includes image-handling functions, largely cloning the image-processing toolbox of MATLAB®. Thus, the images are handled as `numpy` arrays which is convenient, except that the array will not contain any information about the image format used, and sometimes image conversion is done behind the scenes.

The Python imaging library (PIL) provides an `Image` object class which, among other things, knows what image format is being used. Furthermore, `Image` objects can easily be converted to `numpy` arrays as required. The companion software uses `PIL` to load images, because that seems to give a more consistent array representation.

An image can be opened using

```
from PIL import Image
im = Image.open("image.png")
```

The open() method is lazy, so the image will only be loaded when it is necessary. To convert to an array, we need the numpy module, as follows:

import numpy as np
I = np.array(im)

To convert back from an array to an Image object, we can use

im2 = Image.fromarray(I)

To save the image, it is convenient to use PIL:

im2.save("foobar.png")

The array representation makes it easy to show the image using pyplot:

plt.imshow(I)

3.4.5 An Example: Image Histogram

To summarise, Code Example 3.1 presents sample code to generate an image histogram as presented in Figure 2.5, using the Python features discussed above. Readers with some experience of Python will surely find shorter ways to achieve the same result. The purpose of the example is to demonstrate basic functions, which we have found to be robust and flexible, and which will be used in different contexts within the book. New functions are used to compute the histogram and to display bar charts. The syntax should be intuitive and the help function in ipython would explain the rest.

The xrange(N) function returns an iterator over the integers $0, 1, \ldots, N-1$. The histogram function calculates the histogram as an array of frequencies, and the bar function plots this histogram as a bar chart. Further information can be sought using the inline help.

Code Example 3.1 A complete example, generating a histogram plot as presented in Figure 2.5

```
In [1]: import numpy as np
In [2]: import matplotlib.pyplot as plt
In [3]: from PIL import Image
In [4]: im = np.array( Image.open( "gray.tif" ) )
In [5]: bins = [ x - 0.5 for x in xrange(257) ]
In [6]: (h,b) = np.histogram( im, bins )
In [7]: plt.bar( range(256), h )
In [8]: plt.savefig( "histogram.pdf" )
Out[8]:
[...]
```

3.5 Images for Testing

There are many image collections available for people who want to test their image-processing algorithms. In the experiments in this book, we have primarily used the BOWS collection (Ecrypt Network of Excellence, 2008), which has been compiled and published for the benefit of competitions in steganalysis and watermarking. The BOWS collection contains 10 000 never-compressed photographs from different camera sources. It does include raw image data, but we have used 512 × 512 grey-scale images that they provide in PGM format.

For most of the classifier evaluations, we have only used the first 5000 images, half of which have been used as clean images and the other half as cover images for steganograms. The training set is 500 images randomly selected from each class, and 4000 images (or 2000 per class) have been used for testing.

Additionally, we have used a couple of our own photographs to illustrate the behaviour of individual features. We have refrained from using standard images here for copyright reasons. These images are displayed in Figure 3.4. For the

(a)   (b)

(c)   (d)

Figure 3.4 The images used for feature testing: (a) beach; (b) castle; (c) duck; (d) Robben Island

most part, we use a low-resolution grey-scale version. The Robben Island image is 640×480, while the other three are 640×428.

3.6 Further Reading

We will revisit the fundamental disciplines we have discussed later in the book, and readers seeking more depth may just want to read on. For those looking immediately for more dedicated books, the author recommends the following: Theodoridis and Koutroumbas (2009) for classification and machine learning; Lutz (2010) for an introduction to Python; Bhattacharyya and Johnson (1977) for an introduction to statistics. A more advanced and critical text on statistical estimation and inference can be found in MacKay (2003).

The software implementations, pysteg, used in the experiments in this book have been released on the companion website (http://www.ifs.schaathun.net/pysteg/) under the GNU Public Licence. Very recently, Kodovský and Fridrich (2011) have published MATLAB® implementations of a number of feature extractions. Each implementation suite includes many feature vectors which the other is missing.

Part II
Features

4

Histogram Analysis

Analysis of the histogram has been used in steganalysis since the beginning, from the simple pairs-of-values or χ^2 test to analysis of the histogram characteristic function (HCF) and second-order histograms. In this chapter, we will discuss a range of general concepts and techniques using the histogram in the spatial domain. Subsequent chapters will make use of these techniques in other domains.

4.1 Early Histogram Analysis

The histogram was much used in statistical steganalysis in the early days, both in the spatial and the JPEG domains. These techniques were targeted and aimed to exploit specific artifacts caused by specific embedding algorithms. The most well-known example is the pairs-of-values or χ^2 test, which we discussed in Section 2.3.4.

None of the statistical attacks are difficult to counter. For each and every one, new embedding algorithms have emerged, specifically avoiding the artifact detected. Most instructively, Outguess 0.2 introduced so-called *statistics-aware embedding*. Sacrificing some embedding capacity, a certain fraction of the coefficients was reserved for dummy modifications designed only to even out the statistics and prevent the (then) known statistical attacks.

A key advantage of machine learning over statistical attacks is that it does not automatically reveal a statistical model to the designer. The basis of any statistical attack is some theoretical model capturing some difference between steganograms and natural images. Because it is well understood, it allows the steganographer to design the embedding to fit the model of natural images, just like the steganalyst can design the attack to detect discrepancies with the model. Machine learning, relying on brute-force computation, gives a model too complex for direct, human analysis. Thus it deprives the steganalyst of some of the insight which could be used to improve the steganographic embedding.

Machine Learning in Image Steganalysis, First Edition. Hans Georg Schaathun.
© 2012 John Wiley & Sons, Ltd. Published 2012 by John Wiley & Sons, Ltd.

It should be mentioned that more can be learnt from machine learning than what is currently known in the steganalysis literature. One of the motivations for feature selection in Chapter 13 is to learn which features are really significant for the classification and thus get insight into the properties of objects of each class. Further research into feature selection in steganalysis just might provide the insight to design better embedding algorithms, in the same way as insight from statistical steganalysis has improved embedding algorithms in the past.

4.2 Notation

Let I be an image which we view as a matrix of pixels. The pixel value in position (x, y) is denoted as $I[x, y]$. In the case of colour images, $I[x, y]$ will be a tuple, and we denote the value of colour channel k by $I^k[x, y]$. We will occasionally, when compactness is required, write $I_{x,y}$ instead of $I[x, y]$.

The histogram of I is denoted by $h_I(v)$. For a grey-scale image we define it as

$$h_I(v) = \#\{(x, y) \mid I[x, y] = v\}, \tag{4.1}$$

where $\#S$ denotes the cardinality of a set S. For a colour image we define the histogram for each colour channel k as

$$h_I^{(k)}(v) = \#\left\{(x, y, k) \mid I_{x,y}^{(k)} = v\right\}.$$

We will sometimes use a subscript to identify a steganogram I_s or a cover image I_c. For simplicity, we will write h_s and h_c for their histograms, instead of h_{I_s} and h_{I_c}.

We define a function δ for any statement E, such that $\delta(E) = 1$ if E is true and $\delta(E) = 0$ otherwise. We will use this to define joint histograms and other histogram-like functions later. As an example, we could define h_I as

$$h_I(v) = \sum_{x,y} \delta(I[x, y] = v),$$

which is equivalent to (4.1).

4.3 Additive Independent Noise

A large class of stego-systems can be described in a single framework, as adding an independent noise signal to the image signal. Thus we denote the steganogram I_s as

$$I_s = I_c + X,$$

where I_c is the cover object and X is the distortion, or noise, caused by steganographic embedding. As far as images are concerned, I_s, I_c and X are $M \times N$ matrices, but this framework may extend to one-dimensional signals such as audio signals. We say that X is additive independent noise when X is statistically independent of the cover I_c, and each sample $X[i,j]$ is identically and independently distributed.

The two most well-known examples of stego-systems with additive independent noise are spread-spectrum image watermarking (SSIS) and LSB matching. In SSIS (Marvel et al., 1999), this independent noise model is explicit in the algorithm. The message is encoded as a Gaussian signal independently of the cover, and then added to the cover. In LSB matching, the added signal X is statistically independent of the cover, even though the algorithm does calculate X as a function of I_c. If the message bit is 0, the corresponding pixel $I_c[i,j]$ is kept constant if it is even, and changed to become even by adding ± 1 if it is odd. For a message bit equal to one, even pixels are changed while odd pixels are kept constant. The statistical independence is obvious by noting that the message bit is 0 or 1 with 50/50 probability, so that regardless of the pixel value, we have probabilities (25%, 50%, 25%) for $X_{i,j} \in \{-1, 0, +1\}$.

The framework is also commonly used on other embedding algorithms, where X and I_c are statistically dependent. This is, for instance, the case for LSB replacement, where $X_{i,j} \in \{0, +1\}$ if $I_c[i,j]$ is even, and $X_{i,j} \in \{-1, 0\}$ otherwise. Even though X is not independent of I_c, it should not be surprising if it causes detectable artifacts in statistics which are designed to detect additive independent noise. Hence, the features we are going to discuss may very well apply to more stego-systems than those using additive independent noise, but the analysis and justifications of the features will not apply directly.

4.3.1 The Effect of Noise

Many authors have studied the effect of random independent noise. Harmsen and Pearlman (2003) and Harmsen (2003) seem to be the pioneers, and most of the following analysis is based on their work. Harmsen's original attack is probably best considered as a statistical attack, but it is atypical in that it targets a rather broad class of different embedding algorithms. Furthermore, the statistics have subsequently been used in machine learning attacks.

Considering the steganogram $I_s = I_c + X$, it is clear that I_s and I_c have different distributions, so that it is possible, at least in theory, to distinguish statistically between steganograms and clean images. In particular, I_s will have a greater variance, since

$$\mathrm{Var}\, I_s = \mathrm{Var}\, I_c + \mathrm{Var}\, X,$$

and $\mathrm{Var}\, X > 0$. To make the embedding imperceptible, X will need to have very low variance compared to I_c. The objective is to define statistics which can pick up this difference in distribution.

Theorem 4.3.1 *Consider two independent stochastic variables X and Y, with probability mass functions f_X and f_Y respectively. The sum $Z = X + Y$ will then have probability mass function given as the convolution $f_Z = f_X * f_Y$. The same relationship holds for probability density functions.*

The proof follows trivially from the definition of convolution, which we include for completeness.

Definition 4.3.2 (Convolution) *The* convolution *of two functions $f(x)$ and $g(x)$ is defined as*

$$(g * f)(x) = \sum_{t=-\infty}^{\infty} f(t)g(x - t),$$

for functions defined on a discrete domain. If the domain of either function is bounded, the limits of the sum must be restricted accordingly.

The relationship between the histograms is easier to analyse in the frequency domain. This follows from the well-known Convolution Theorem (below), which is valid in the discrete as well as the continuous case.

Theorem 4.3.3 (Convolution Theorem) *Given two functions $f(x)$ and $g(x)$, and their Fourier transforms $F(\omega)$ and $G(\omega)$, if $h(x) = (f * g)(x)$ then the Fourier transform of $h(x)$ is given as $H(\omega) = F(\omega) \cdot G(\omega)$ for each ω.*

Theorem 4.3.1 tells us that the addition of independent noise acts as a filter on the histogram. In most cases, the added noise is Gaussian or uniform, and its probability mass function (PMF) forms a low-pass filter, so that it will smooth high-frequency components in the PMF while keeping the low-frequency structure. Plotting the PMFs, the steganogram will then have a smoother curve than the cover image.

Obviously, we cannot observe the PMF or probability distribution function (PDF), so we cannot apply Theorem 4.3.1 directly. The histogram is an empirical estimate for the PMF, as the expected value of the histogram is $E(h(x)) = P(X = x)$. Many authors have argued that Theorem 4.3.1 will hold approximately when the histogram $h(x)$ is substituted for $f(x)$. The validity of this approximation depends on a number of factors, including the size of the image. We will discuss this later, when we have seen a few examples.

In Figure 4.1, we see the histogram of a cover image and steganograms created at full capacity with LSB and LSB±. It is not very easy to see, but there is evidence, especially in the range $50 \leq x \leq 100$, that some raggedness in the cover image is smoothed by the additive noise.

It is well known that never-compressed images will tend to have more ragged histograms than images which have been compressed, for instance with JPEG. However, as our never-compressed sample shows, the raggedness is not necessarily very pronounced. Harmsen (2003) had examples where the histogram was considerably more ragged, and the smoothing effect of the noise became more pronounced. Evidently, he must have used different image sources; in the sample we studied, the cover images tended to have even smoother histograms than this one.

The smoothing effect can only be expected to be visible in large images. This is due to the laws of large numbers, implying that for sufficiently large samples, the

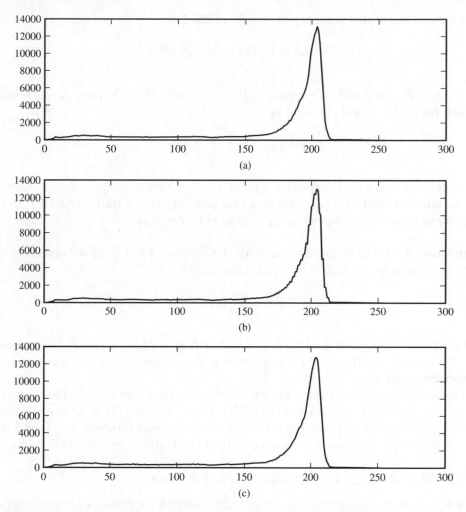

Figure 4.1 Sample histograms for clean images and steganograms, based on the 'beach' image: (a) cover image; (b) LSB replacement; (c) LSB matching

histograms are good approximations for the PMF, and Theorem 4.3.1 gives a good description of the histograms. For small images, the histogram will be ragged because of the randomness of the small sample, both for the steganogram and for the cover. In such a case one will observe a massive difference between the actual histogram of the steganogram and the histogram estimated using Theorem 4.3.1.

4.3.2 The Histogram Characteristic Function

The characteristic function (CF) is the common transform for studying probability distributions in the frequency domain. It is closely related to the Fourier transform.

The CF of a stochastic variable X is usually defined as

$$\phi(u) = E[e^{iuX}] = \sum_x f_X(x)e^{iux},$$

where f_X is the PMF and i is the imaginary unit solving $x^2 = -1$. The discrete Fourier transform (DFT), however would be

$$F(u) = E[e^{-iuX}] = \sum_x f_X(x)e^{-iux}.$$

The relevant theoretical properties are not affected by this change of sign. Note in particular that $|\phi(u)| = |F(u)|$. One may also note that $\phi(u)$ is the inverse DFT. In the steganalysis literature, the following definition is common.

Definition 4.3.4 *The* histogram characteristic function $H(\omega)$ *is defined as the discrete Fourier transform of the histogram $h(x)$. In other words,*

$$H(\omega) = \sum_x h(x)e^{-i\omega x}.$$

Combining Theorems 4.3.1 and 4.3.3, we note that $H_s(\omega) = H_c(\omega) \cdot H_x(\omega)$ where H_s, H_c and H_x are the HCF of, respectively, the steganogram, the cover and the embedding distortion.

Theoretical support for using the HCF rather than the histogram in the context of steganalysis was given by Xuan *et al.* (2005a). They compared the effect of embedding on the PDF and on its Fourier transform or characteristic function, and found that the characteristic function was more sensitive to embedding than the PDF itself.

4.3.3 Moments of the Characteristic Function

Many authors have used the so-called HCF moments as features for steganalysis. The concept seems to have been introduced first by Harmsen (2003), to capture the changes in the histogram caused by random additive noise. The following definition is commonly used.

Definition 4.3.5 *The nth-order moment of the HCF is defined as*

$$m_n = \frac{\sum_{k=0}^{\lfloor N/2 \rfloor} \left(\frac{k}{N}\right)^n |H(k)|}{\sum_{k=0}^{\lfloor N/2 \rfloor} |H(k)|}, \tag{4.2}$$

where H is the HCF and N is its length. The first-order moment of the HCF is also known as the HCF centre of mass or HCF-COM.

The use of the term 'moment' in steganalysis may be confusing. We will later encounter statistical moments, and the moments of a probability distribution are also well known. The characteristic function is often used to calculate the moments of the corresponding PDF. However, the HCF moments as defined above seem to be unrelated to the more well-known concept of moments. On the contrary, if we replace H by the histogram h in the definition of m_n, we get the statistical moments of the pixel values.

Because the histogram $h(x)$ is real-valued, $|H(k)|$ is symmetric, making the negative frequencies redundant, which is why we omit them in the sums. Most authors include the zero frequency $H(0)$ in the sum, but Shi $et\ al.$ (2005a) argue that, because this is just the total number of pixels in the image, it is unaffected by embedding and may make the moment less sensitive.

Python Tip 4.1 *The indexing of the DFT signal in (4.2) is consistent with the indexing in Python, when* `H[k]` *is returned by* `numpy.fft.fft`. *The zero frequency is* `H[0]`. *The positive frequencies are* `H[1:(floor(N/2)+1)]`, *and the negative frequencies are* `H[ceil(N/2):]`. *When N is even,* `H[N/2]` *is both the positive and negative Nyquist frequency. Thus, we would sum over the elements of* `H[:(floor(N/2)+1)]` *in the calculation of a HCF moment.*

The HCF-COM m_1 was introduced as a feature by Harmsen (2003), and has been used by many authors since. We will study it in more detail. Code Example 4.1 shows how it can be calculated in Python, using the FFT and histogram functions from numpy. Let $C(I) = m_1$ denote the HCF-COM of an arbitrary image I. The key to its usefulness is the following theorem.

Theorem 4.3.6 (Harmsen and Pearlman, 2003) *For an embedding scheme with non-increasing* $|H_X(k)|$ *for* $k = 0, 1, \ldots, N - 1$, *the HCF-COM decreases or remains the same after embedding, i.e.*

$$C(I_s) \leq C(I_c),$$

with equality if and only if $|H_X(k)| = 1$ *for all k.*

Code Example 4.1 Python function to calculate the HCF moments of an image `im`

```
def hcfmoments(im,order=1):
    """Calculate the HCF moments of the
    given orders for the image im."""
    h = np.histogram( h, range(257) )
    H0 = np.fft.fft(h)
    H = np.abs( H0[1:129] )
    M = sum( [ (float(k)/(256))**order*H[k] for k in xrange(128) ] )
    S = np.sum(H)
    return M / S
```

The proof depends on the following version of Čebyšev's[1] inequality. Because this is not the version most commonly found in the literature, we include it with a proof based on Hardy *et al.* (1934).

Lemma 4.3.7 *If p_i is a non-negative sequence, and a_i and b_i are sequences where one is non-increasing and the other is non-decreasing, then we have that*

$$\sum_{i=1}^{n} p_i \sum_{j=1}^{n} p_j a_j b_j \leq \sum_{i=1}^{n} p_i a_i \sum_{j=1}^{n} p_j b_j.$$

If a_i and b_i are either both non-increasing or both non-decreasing, then the inequality is reversed.

Proof. We consider the difference between the left- and right-hand side of the inequality, and write

$$D = \sum_{i=1}^{n} p_i a_i \sum_{j=1}^{n} p_j b_j - \sum_{i=1}^{n} p_i \sum_{j=1}^{n} p_j a_j b_j$$

$$= \sum_{i=1}^{n} \sum_{j=1}^{n} p_i p_j (a_i b_j - a_j b_j).$$

Note that for any index pair (i, j) included in the sum, (j, i) is also included. Therefore, we can write

$$D = \sum_{i=1}^{n} \sum_{j=1}^{n} p_i p_j \frac{1}{2} (a_i b_j - a_j b_j + a_j b_i - a_i b_i)$$

$$= \frac{1}{2} \sum_{i=1}^{n} \sum_{j=1}^{n} p_i p_j (a_i - a_j)(b_j - b_i).$$

Since a_i and b_i are oppositely ordered, $a_i - a_j$ and $b_j - b_i$ have the same sign. With $p_i \geq 0$, we thus get that $D \geq 0$. If, on the contrary, a_i and b_i are similarly ordered, we get $D \leq 0$.

Proof of Theorem 4.3.6. Since $|H_X(k)|$ is non-increasing in k and $|H_c(k)|$ is non-negative for all k, we can use $a_k = k$, $b_k = |H_X(k)|$ and $p_k = |H_c(k)|$ in Lemma 4.3.7 to get that

$$\sum_{k \in \mathcal{K}} |H_c(k)| \sum_{k \in \mathcal{K}} k |H_X(k)| \cdot |H_c(k)| \leq \sum_{k \in \mathcal{K}} k |H_c(k)| \sum_{k \in \mathcal{K}} |H_X(k)| \cdot |H_c(k)|. \qquad (4.3)$$

---

[1] Also spelt Chebyshev or Tchebychef.

Rearranging terms, we get

$$\frac{\sum_{k\in\mathcal{K}} k \left|H_X(k)\right| \left|H_c(k)\right|}{\sum_{k\in\mathcal{K}} \left|H_X(k)\right| \left|H_c(k)\right|} \leq \frac{\sum_{k\in\mathcal{K}} k \left|H_c(k)\right|}{\sum_{k\in\mathcal{K}} \left|H_c(k)\right|}. \tag{4.4}$$

It is easy to see that equality holds in (4.3), and hence in (4.4), if $\left|H_X(k)\right|$ is constant in k. Then, note that $H_X(k = 0) = 1$ because H_X is the Fourier transform of a PMF which has to sum to 1. Since $H_X(k)$ is also non-increasing, we note that equality can only hold if $\left|H_X(k)\right| = 1$ for all k.

In Figure 4.2 we can see the empirical HCF-COM plotted as a function of the length of the embedded message, using LSB matching for embedding. On the one hand, this confirms Theorem 4.3.6; HCF-COM is smaller for steganograms than it is for the corresponding cover image. On the other hand, the difference caused by embedding is much smaller than the difference between different cover images. Thus, we cannot expect HCF-COM to give a working classifier as a sole feature, but it can quite possibly be useful in combination with others.

We note that the high-resolution image has a much smoother and more consistently declining HCF-COM. This is probably an effect of the laws of large numbers; the high-resolution image has more samples and thus the random variation becomes negligible.

We have also made a similar plot using LSB replacement in Figure 4.3. Notably, HCF-COM increases with the message length for LSB replacement. This may come as a surprise, but it demonstrates that the assumption of random *independent* noise is critical for the proof of Theorem 4.3.6. In contrast, the effect of LSB replacement on HCF-COM seems to be greater than that of LSB matching, so it is not impossible that HCF-COM may prove useful as a feature against LSB replacement.

As we shall see throughout the book, the HCF moments have been used in many variation for steganalysis, including different image representations and higher-order moments as well as HCF-COM.

4.3.4 Amplitude of Local Extrema

Zhang *et al.* (2007) proposed a more straight forward approach to detect the smoothing effect of additive noise. Clean images very often have narrow peaks in the histogram, and a low-pass filter will tend to have a very pronounced effect on such peaks. Consider, for instance, a pixel value x which is twice as frequent as either neighbour $x \pm 1$. If we embed with LSB matching at full capacity, half of the x pixels will change by ± 1. Only a quarter of the $x \pm 1$ pixels will change to exactly x, and this is only half as many. Hence, the peak will be, on average, 25% lower after embedding.

In Zhang *et al.*'s (2007) vocabulary, these peaks are called local extrema; they are extreme values of the histogram compared to their local neighbourhood. They proposed a way to measure the amplitude of the local extrema (ALE). In symbolic notation, the local extrema is the set $\mathcal{E}$ defined as

$$\mathcal{E} = \{x \mid (h(x) - h(x-1))(h(x) - h(x+1)) > 0\}. \tag{4.5}$$

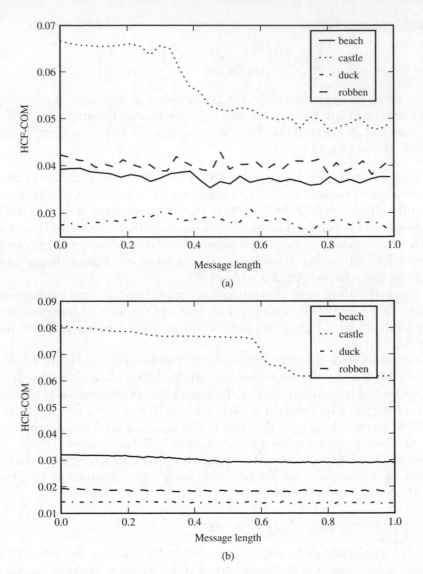

Figure 4.2 Plot of HCF-COM for different test images and varying message lengths embedded with LSB matching. We show plots for both the high-resolution originals: (a) and the low-resolution grey-scale images; (b) that we use for most of the tests

For 8-bit images, $\mathcal{E} \subset \{1, 2, \ldots, 254\}$, as the condition in (4.5) would be undefined for $x = 0, 255$. The basic ALE feature, introduced by Zhang *et al.* (2007), is defined as

$$f_0 = \sum_{x \in \mathcal{E}} \left| 2h(x) - h(x-1) - h(x+1) \right|. \tag{4.6}$$

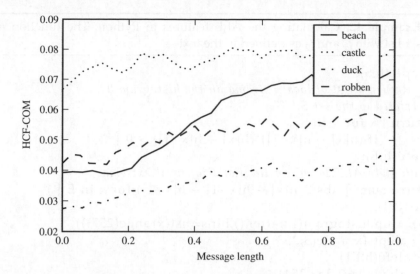

Figure 4.3 Plot of HCF-COM for varying message lengths with LSB replacement

The ragged histogram is typical for images that have never been compressed, and the experiments reported by Zhang *et al.* (2007) show that f_0 is very effective for never-compressed images, but considerably less discriminant for images that have ever been JPEG compressed.

Irregular effects occur in the histogram close to the end points. The modification of LSB matching is ±1 with uniform distribution, except at 0 and 255. A zero can only change to one, and 255 only to 254. Because of this, Cancelli *et al.* (2008) proposed, assuming 8-bit grey-scale images, to restrict $\mathcal{E} \subset \{3,4,\ldots,252\}$. The resulting modified version of f_0 thereby ignores values which are subject to the exception occurring at 0 and 255. To avoid discarding the information in the histogram at values $1,2,253,254$, they define a second feature on these points. Thus, we get two first-order ALE features as follows:

$$f_1 = \sum_{x\in\mathcal{E}\cap\{3,4,\ldots,252\}} \left|2h(x) - h(x-1) - h(x+1)\right|,$$

$$f_2 = \sum_{x\in\mathcal{E}\cap\{1,2,253,254\}} \left|2h(x) - h(x-1) - h(x+1)\right|.$$

Python definitions of these features are shown in Code Example 4.2. In Figure 4.4 we can see how f_1 depends on the embedded message length for a couple of sample images. The complete set of ALE features (ALE-10) also includes second-order features, which we will discuss in Section 4.4.2.

Code Example 4.2 Calculating the ALE features in Python. The function `ale1d` returns a list with f_1 and f_2 as defined in the text

```
def alepts(h,S):
    """Return a list of local extrema on the histogram h
    restricted to the set S."""
    return [ x for x in S
                if (h[x] − h[x−1])*(h[x] − h[x+1]) > 0 ]
def alef0(h,E):
    "Calculate_ALE_based_on_the_list_E_of_local_extrema."
    return sum( [ abs(2*h[x] − h[x−1] − h[x+1]) for x in E ] )
def ale1d(I):
    (h,b) = np.histogram(I.flatten(),bins=list(xrange(257)))
    E1 = alepts(xrange(3,253))
    f1 = alef0(h,E1)
    E2 = alepts([1,2,253,254])
    f2 = alef0(h,E2)
    return [f1,f2]
```

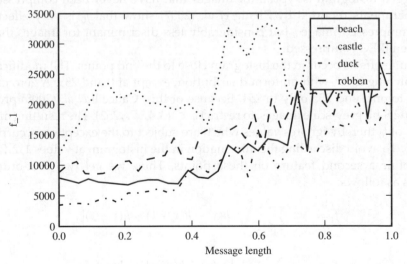

Figure 4.4 The ALE feature f_1 as a function of the message length for a few sample images with LSB replacement. The message length is given as a fraction of capacity

4.4 Multi-dimensional Histograms

So far, we have discussed HCF-COM as if we were going to apply it directly to grey-scale images, but the empirical tests in Figures 4.2 and 4.3 show that this is unlikely to give a good classifier. In order to get good accuracy, we need other features. There are indeed many different ways to make effective use of the HCF-COM feature, and we will see some in this section and more in Chapter 9.

Theorem 4.3.1 applies to multi-dimensional distributions as well as one-dimensional ones. Multi-dimensional distribution applies to images in at least two different ways. Harmsen and Pearlman (2003) considered colour images and the joint distribution of the colour components in each pixel. This gives a three-dimensional distribution and a 3-D (joint) histogram. A more recent approach, which is also applicable to grey-scale images, is to study the joint distribution of adjacent pixels.

The multi-dimensional histograms capture correlation which is ignored by the usual one-dimensional histogram. Since pixel values are locally dependent, while the noise signal is not, the multi-dimensional histograms will tend to give stronger classifiers. Features based on the 1-D histogram are known as first-order statistics. When we consider the joint distribution of n pixels, we get nth-order statistics.

It is relatively easy to build statistical models predicting the first-order features, and this enables the steganographer to adapt the embedding to keep the first-order statistics constant under embedding. Such *statistics-aware* embedding was invented very early, with Outguess 0.2. Higher-order models rapidly become more complex, and thus higher-order statistics (i.e. $n > 1$) are harder to circumvent.

4.4.1 HCF Features for Colour Images

Considering a colour image, there is almost always very strong dependency between the different colour channels. Look, for instance, at the scatter plots in Figure 4.5. Each point in the plot is a pixel, and its coordinates (x, y) represent the value in two different colour channels. If the channels had been independent, we would expect an

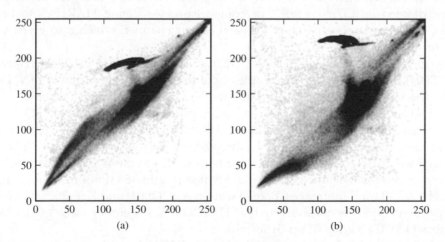

(a)   (b)

Figure 4.5 Scatter plots showing the correlation between two colour channels: (a) red/green; (b) red/blue. The plots are based on the 'castle' image, scaled down to 512 × 343

almost evenly grey plot. As it is, most of the possible combinations hardly ever occur. This structure in the joint distribution is much more obvious than the structure in the one-dimensional histogram. Thus, stronger classifiers can be expected.

The joint distribution of pixel values across three colour channels can be described as a three-dimensional (joint) histogram

$$h_I(u, v, w) = \#\{(x, y) \mid I^R(x, y) = u \wedge I^G(x, y) = v \wedge I^B(x, y) = w\}.$$

The only difference between this definition and the usual histogram is that the 3-D histogram is a function of three values (u, v, w) instead of one. Theorem 4.3.6 can easily be generalised for joint probability distributions, as a multi-dimensional convolution. Thus, the theory we developed in the 1-D case should apply in the multi-dimensional case as well.

The multi-dimensional HCF is a multi-dimensional Fourier transform of the joint histogram, and the HCF centre of mass can be generalised as follows.

Definition 4.4.1 *The generalised centre of mass of a d-dimensional HCF is a d-tuple given by*

$$C(H(\mathbf{k})) = [c_1, c_2, \ldots, c_d], \quad where$$

$$c_i = \frac{\sum_{\mathbf{k}} k_i \, |H(\mathbf{k})|}{\sum_{\mathbf{k}} |H(\mathbf{k})|},$$

and $\mathbf{k} = (k_1, \ldots, k_d)$. *Each element* c_i *of* $C(H(\mathbf{k}))$ *is sometimes called a marginal moment of order 1.*

Having generalised all the concepts for multi-dimensional distributions, we let $H^3(k, l, m)$ denote the HCF obtained as the 3-D DFT of $h_I(u, v, w)$. The feature vector of Harmsen and Pearlman (2003), which we denote as HAR3D-3, is a three-dimensional vector defined as $C(H^3(k, l, m))$. For comparison, we define HAR1D-3 to be the HCF-COM calculated separately for each colour channel, which also gives a 3-D feature vector.

The disadvantage of HAR3D-3 is that it is much slower to calculate (Harmsen *et al.*, 2004) than HAR1D-3. The complexity of the d-D DFT algorithm is $O(N^d \log N)$ where the input is an array of N^d elements. Thus, for 8-bit images, where $N = 256$, HAR1D-3 is roughly 20 000 times faster to compute compared with HAR3D-3.

A compromise between HAR1D-3 and HAR3D-3 is to use the 2-D HCF for each pair of colour channels. For each of the three 2-D HCFs, the centre of mass $C(H(k_1, k_2))$ is calculated. This gives two features for each of three colour pairs. The result is a 6-D feature vector which we call HAR2D-6. It is roughly 85 times faster to compute than HAR3D-3 for 8-bit images. Both Harmsen *et al.* (2004) and Ker (2005a) have run independent experiments showing that HAR2D-6 has accuracy close to HAR3D-3, whereas HAR1D-3 is significantly inferior.

Code Example 4.3 shows how to calculate HAR2D-6: `har2d` calculates the feature vector from an image, using `hdd()` to calculate the HCF centre of mass from a given pair of colour channels, which in turn uses `hcom` to calculate the centre of mass for an arbitrary HCF.

Code Example 4.3 Functions to calculate HAR2D-6 from an image

---

```
def har2d(I):
  "Return_the_HAR2D−6_feature_vector_of_I_as_a_list."
  (R,G,B) = (I[:,:,0],I[:,:,1],I[:,:,2])
  return hdd((R,B)) + hdd((R,G)) + hdd((B,G))
def hdd(L):
  """

  Given a list L of 8−bit integer arrays, return the HCF−COM
  based on the joint histogram of the elements of L.
  """

  bins = tuple([ list(xrange(257)) for i in L ])
  L = [ X.flatten() for X in L ]
  (h,b) = np.histogramdd( L, bins=bins )
  H = np.fft.fftn(h)
  return hcom(H)
def hcom(H1):
  "Calculate_the_centre_of_mass_of_a_2−D_HCF_H."
  H = H1[0:129,0:129]
  S = np.sum(H)
  return [ hcfm( np.sum(H,1) )/S, hcfm( np.sum(H,0) )/S ]
```

4.4.2 The Co-occurrence Matrix

For grey-scale images the previous section offers no solution, and we have to seek higher-order features within a single colour channel. The first-order features consider pixels as if they were independent. In reality, neighbouring pixels are highly correlated, just like the different colour channels are. The scatter plot in Figure 4.6 is analogous to those showing colour channels. We have plotted all the horizontal pixel pairs $(x,y) = (I[i,j], I[i,j+1])$ of a grey-scale image. Dark colour indicates a high density of pairs, and as we can see, most of the pairs cluster around the diagonal where $I[i,j] \approx I[i,j+1]$. This is typical, and most images will have similar structure. Clearly, we could consider larger groups of pixels, but the computational cost would then be significant.

Numerically, we can capture the distribution of pixel pairs in the *co-occurrence matrix*. It is defined as the matrix $M_{\delta_x,\delta_y} = [m_{v,w}]$, where

$$M_{\delta_x,\delta_y}[v,w] = m_{v,w} = \# \left\{ (x,y) \mid I[x,y] = v, I[x+\delta_x, y+\delta_y] = w \right\}.$$

We are typically interested in the co-occurrence matrices for adjacent pairs, that is $M_{i,j}$ for $i,j = -1, 0, +1$, but there are also known attacks using pixels at distance greater than 1. The co-occurrence matrix is also known as the second-order or 2-D histogram. An example of how to code it in Python is given in Code Example 4.4.

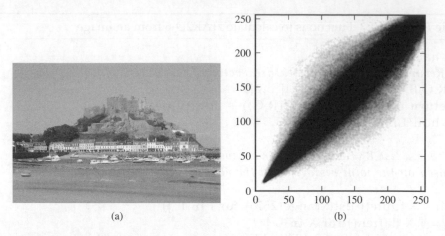

(a)   (b)

Figure 4.6   Image (a) with its horizontal co-occurrence scatter diagram (b)

Code Example 4.4 Calculate the co-occurrence matrix M_{d_1,d_2} of C in the given direction d1 , d2

```
def cooccurrence(C,(d1,d2)):
    (C1,C2) = splitMatrix(C,(d1,d2))
    (h,b) = np.histogram2d( C1.flatten(),C2.flatten(),
        bins=range(0,257 )
    return h
def splitMatrix(C,(d1,d2)):
    """Return the two submatrices of C as needed to calculate
    co-occurrence matrices."""
    return (sm(C,d1,d2),sm(C,−d1,−d2))
def sm(C,d1,d2):
    "Auxiliary function for splitMatrix()."
    (m,n) = C.shape
    (m1,m2) = (max(0,−d1), min(m,m−d1))
    (n1,n2) = (max(0,−d2), min(n,n−d2))
    return C[m1:m2,n1:n2]
```

Sullivan's Features

Additive data embedding would tend to make the distribution more scattered, with more pairs further from the diagonal in the scatter plot. This effect is exactly the same as the one we discussed with respect to the 1-D histogram, with the low-pass filter caused by the noise blurring the plot. The co-occurrence matrix M_s of a steganogram will be the convolution $M_c * M_X$ of the co-occurrence matrices M_c and M_X of the cover image and the embedding distortion. The scattering effect becomes very prominent if the magnitude of the embedding is large. However, with LSB embedding, and other techniques adding ± 1 only, the effect is hardly visible in the scatter plot, but may be detectable statistically or using machine learning.

Sullivan *et al.* (2005) suggested a simple feature vector to capture the spread from the diagonal. They used a co-occurrence matrix $M = [m_{i,j}]$ of adjacent pairs from a linearised (1-D) sequence of pixels, without addressing whether it should be parsed row-wise or column-wise from the pixmap. This, however, is unlikely to matter, and whether we use $M = M_{1,0}$ or $M = M_{0,1}$, we should expect very similar results.

For an 8-bit grey-scale image, M has 2^{16} elements, which is computationally very expensive. Additionally, M may tend to be sparse, with most positive values clustering around the main diagonal. To address this problem, Sullivan *et al.* (2005) used the six largest values $m_{k,k}$ on the main diagonal of M, and for each of them, a further ten adjacent entries

$$m_{k,k}, m_{k,k-1}, m_{k,k-2}, \ldots, m_{k,k-10},$$

with the convention that $m_{k,l} = 0$ when the indices go out of bounds, that is for $l < 0$. This gives 66 features.

Additionally, they used every fourth of the values on the main diagonal, that is

$$m_{k,k} \quad \text{for } k \equiv 1 \ (\text{mod } 4).$$

This gives another 64 features, and a total of 130. We will denote the feature vector as SMCM-130.

The SMCM-130 feature vector was originally designed to attack a version of SSIS of Marvel *et al.* (1999). SSIS uses standard techniques from communications and error-control coding to transmit the steganographic image while treating the cover image as noise. The idea is that the embedding noise can be tailored to mimic naturally occurring noise, such as sensor noise in images. The theory behind SSIS assumes floating-point pixel values. When the pixels are quantised as 8-bit images, the variance of the Gaussian embedding noise has to be very large for the message to survive the rounding errors, typically too large to mimic naturally occurring noise. In the experiments of Marvel *et al.* using 8-bit images, SSIS becomes very easily detectable. Standard LSB replacement or matching cause much less embedding noise and are thus harder to detect. It is possible that SSIS can be made less detectable by using raw images with 12- or 14-bit pixels, but this has not been tested as far as we know.

The 2-D HCF

To study the co-occurrence matrix in the frequency domain, we can follow the same procedure as we used for joint histograms across colour channels. It applies to any co-occurrence matrix M. The 2-D HCF $H^2(k, l)$ is obtained as the 2-D DFT of M.

Obviously, we can form features using the centre of mass from Definition 4.4.1. However, Ker (2005b) used a simpler definition to get a scalar centre of mass from

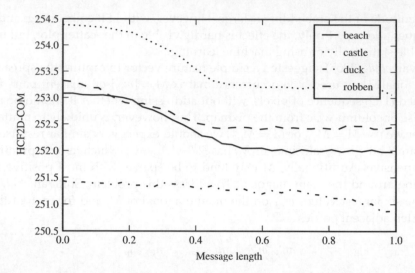

Figure 4.7 Ker's 2-D HCF-COM feature as a function of embedded message length for LSB matching

a 2-D HCF. He considered just the sum of the pixels in the pair, and defined a centre of mass as

$$C_2(M) = \frac{\sum_{i,j=0}^{n}(i+j)\left|H_M^2(i,j)\right|}{\sum_{i,j=0}^{n}\left|H_M^2(i,j)\right|},\qquad(4.7)$$

where H_M^2 is the 2-D HCF of the co-occurrence matrix M. Obviously, this idea may be extended to use higher-order moments of the sum $(i+j)$ as well. Ker (2005b) used just $C_2(M_{0,1})$ as a single discriminant, and we are not aware of other authors who have used 2-D HCFs in this way.

Comparing Figure 4.7 to Figure 4.2, we see that Ker's 2-D HCF-COM feature varies much more with the embedded message length, and must be expected to give a better classifier.

Amplitude of Local Extrema

The ALE features can be generalised for joint distributions in the same way as the HCF moments. The second-order ALE features are calculated from the co-occurrence matrices, or 2-D adjacency histograms in the terminology of Cancelli *et al.* (2008).

A local extremum in a 1-D histogram is a point which is either higher or lower than both neighbours. In 2-D, we define a local extremum to be a point where either all four neighbours (above, below, left, right) are higher, or all four are lower. Formally, a local extremum in a co-occurrence matrix M is a point (x,y) satisfying either

$$\forall(x',y') \in \mathcal{N},\ M[x,y] - M[x+x',y+y'] < 0,\quad \text{or}\qquad(4.8)$$

$$\forall (x',y') \in \mathcal{N}, \ M[x,y] - M[x+x',y+y'] > 0, \quad \text{where} \tag{4.9}$$

$$\mathcal{N} = \{(0,1),(1,0),(-1,0),(0,-1)\}.$$

Let $\mathcal{E}_M^*$ be the set of all the local extrema.

A potential problem is a large number of local extrema with only negligible differences in (4.8) or (4.9). Such marginal extrema are not very robust to embedding changes, and ideally we want some robustness criterion to exclude marginal extrema.

It is reasonable to expect the co-occurrence matrix to be roughly symmetric, i.e. $M[x,y] \approx M[y,x]$. After all, $M[y,x]$ would be the co-occurrence matrix of the image flipped around some axis, and the flipped image should normally have very similar statistics to the original image. Thus, a robust local extremum (x,y) should be extremal both with respect to $M[x,y]$ and $M[y,x]$. In other words, we define a set $\mathcal{E}_M$ of robust local extrema, as

$$\mathcal{E}_M = \{(x,y) | (x,y) \in \mathcal{E}_M^* \wedge (y,x) \in \mathcal{E}_M^*\}.$$

According to Cancelli $et\ al.$, this definition improves the discriminating power of the features. Thus, we get a 2-D ALE feature $A(M)$ for each co-occurrence matrix M, defined as follows:

$$A(M) = \sum_{\mathbf{p} \in \mathcal{E}_M} \left| 4M(\mathbf{p}) - \sum_{\mathbf{n} \in \mathcal{N}} M(\mathbf{p} + \mathbf{n}) \right|. \tag{4.10}$$

We use the co-occurrence of all adjacent pairs, horizontally, vertically and diagonally. In other words,

$$M \in \{M_{0,1}, M_{1,0}, M_{1,1}, M_{1,-1}\},$$

to get four features $A(M)$. Figure 4.8 shows how a second-order ALE feature responds to varying message length.

The last of the ALE features are based on the ideas discussed in Section 4.4.2, and count the number of pixel pairs on the main diagonal of the co-occurrence matrix:

$$d(M) = \sum_{k=0}^{255} M[k,k]. \tag{4.11}$$

When the embedding increases the spreading off the main diagonal as discussed previously, $d(M)$ will decrease.

Code Example 4.5 demonstrates how to calculate ALE features in Python. We define the list L as the four different shifts of I by one pixel up, down, left or right. The Boolean array T is first set to flag those elements of I that are smaller than all their four neighbours, and subsequently or'ed with those elements smaller than the neighbours. Thus, T identifies local extrema. Finally, T is and'ed with its transpose to filter out robust local extrema only. The two features A and d are set using the definitions (4.10) and (4.11).

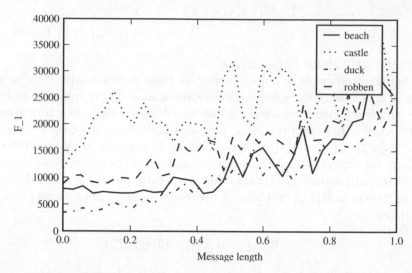

Figure 4.8 The second-order ALE feature $A(M_{0,1})$ as a function of the embedded message length for our test images

---

Code Example 4.5 Calculate the 2-D ALE features $A(M_{d_1,d_2})$ and $d(M_{d_1,d_2})$ for the given direction (d_1, d_2)

```
def ale2d(I,(d1,d2)):
    h = cooccurrence(I,(d1,d2))
    Y = h[1:-1,1:-1]
    L = [h[:-2,1:-1],h[2:,1:-1],h[1:-1,:-2],h[1:-1,2:]]
    T  = reduce( andf, [ ( Y < x ) for x in L ] )
    T |= reduce( andf, [ ( Y > x ) for x in L ] )
    T &= T.transpose()
    A = sum( np.abs(4*Y[T] - sum( [ x[T] for x in L ] )) )
    d = sum( [ h[k,k] for k in xrange(h.shape[0]) ] )
    return [A,d]
```

Table 4.1 Comparison of accuracies of feature vectors discussed in this chapter

Stego-system	LSB	LSB±	LSB	LSB±	LSB	LSB±
Message length	512B		40%		full length	
SMCM-130	51.7%	50.8%	74.1%	77.2%	91.7%	89.8%
ALE-10	50.1%	50.3%	79.1%	79.2%	95.7%	91.3%

Cancelli *et al.* (2008) proposed a 10-D feature vector, ALE-10, comprising the first-order features f_1 and f_2 and eight second-order features. Four co-occurrence matrices $M = M_{0,1}, M_{1,0}, M_{1,1}, M_{1,-1}$ are used, and the two features $A(M)$ and $d(M)$ are calculated for each M, to give the eight second-order features.

4.5 Experiment and Comparison

We will end each chapter in this part of the book with a quick comparison of accuracies on grey-scale images, between feature vectors covered so far. Table 4.1 shows the performance of ALE-10 and SMCM-130. We have not considered colour image or single features in this test. Neither of the feature vectors is able to detect very short messages, but both have some discriminatory power for longer messages.

The SMCM features do badly in the experiments, especially considering the fact that they have high dimensionality. This makes sense as the SMCM features consider the diagonal elements of the co-occurrence matrix separately, where each element would be expected to convey pretty much the same information.

4.5 Experiment and Comparison

5

Bit-plane Analysis

The study of individual bit planes is one of the most classic approaches to steganalysis. By bit plane i we mean the binary image which is obtained by taking only bit i from the binary representation of each pixel. Thus, the eighth bit plane of an 8-bit image is the pixmap reduced modulo two, also known as the LSB plane.

We saw the first example of bit-plane analysis, in the context of visual steganalysis, in Figure 2.4. In this chapter we shall see that bit-plane analysis can also form the basis for features to be used in machine learning. However, to understand the artifacts the features are designed to capture, it is helpful to start with more visual steganalysis.

5.1 Visual Steganalysis

The visual attack from Figure 2.4 is very well known in the literature. It may therefore be surprising that there has been little discussion of when it is appropriate and when it is not. A novice trying to repeat the experiments of classic authors will often struggle. Far from all images will show the nice structure which is so evident in Figure 2.4.

Figure 5.1 shows another example. The image is taken with the same camera, and converted and embedded in the exact same way as the image in Figure 2.4. Some structure is seen in the LSB plane of the palette image, but not as much as in the previous example. In 8-bit grey scale the LSB plane looks completely random. In the steganogram, a random message is embedded in the left-hand half of the image, but this cannot be distinguished visually from the clean image.

It is evident that visual steganalysis of the LSB plane works on only some images. It tends to work more often on palette images than on grey-scale images. This is due to the effect of quantisation working somewhat differently in grey-scale and palette images. In a colour image, the pixel colour is a point in $\mathbb{R}^3$. In the grey-scale image, the colour is first projected down to $\mathbb{R}^1$. Similar colours will become similar tones

---

Machine Learning in Image Steganalysis, First Edition. Hans Georg Schaathun.

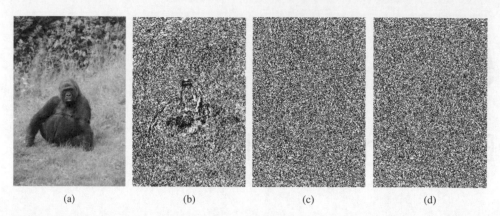

(a)   (b)   (c)   (d)

Figure 5.1 An image with LSB bit plane in different formats: (a) image; (b) palette image LSB; (c) grey-scale image LSB; (d) LSB steganogram LSB. Photo by Welie Schaathun © 2009

of grey. Since much of the image often consists of smooth colour transitions, there will often be a difference of ±1 between adjacent grey-scale pixels, which contributes to the random-looking LSB plane. The palette image has to quantise the colour points directly in the three-dimensional space. This gives a very sparse set of palette colours in 3-D, and a correspondingly serious quantisation error. Colours which differ by ±1 in grey scale may correspond to the same palette colour, which may give monochrome areas in the palette image LSB plane, where the grey-scale LSB plane looks random. Furthermore, palette colours which are closest neighbours in $\mathbb{R}^3$ are not necessarily represented as adjacent 8-bit integers. Whereas similar grey-scale tones differ in the LSB, similar palette colours are equally likely to have the same LSB value.

When visual steganalysis does work, it will often reveal additional information about the embedded message. The message in Figure 5.1 is a random, unbiased bit string. The random-looking pattern is indeed random. In Figure 5.2, we have instead embedded a text message. When the embedding is in consecutive pixels, we can clearly see the structure of the message. The text consists of characters represented as

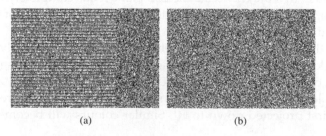

(a)   (b)

Figure 5.2 LSB embedding of a text message: (a) consecutive pixels; (b) random pixels. Image as in Figure 5.1

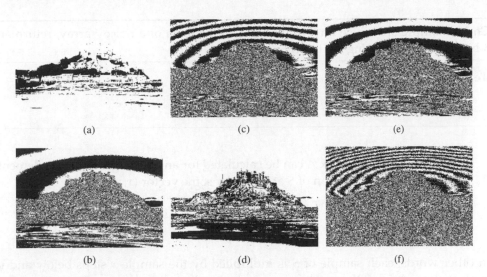

Figure 5.3 Various bit planes of the image from Figure 2.4: (a) bit plane 1; (b) bit plane 3; (c) bit plane 5; (d) bit plane 6; (e) bit plane 7; (f) bit plane 8

integers in $\{0, \ldots, 255\}$. Almost all the characters are English-language letters, which lie in the range $\{65, \ldots, 122\}$. Hence, the first bit of each character is almost always 0, and the second bit almost always 1. In the embedded steganogram, this means that almost every eighth bit is white and subsequent bit is black, leading to the horizontal stripes visible in Figure 5.2(a). In Figure 5.2(b), we have embedded the message bits in random locations, and hence the visual structure disappears.

It is sometimes useful to investigate more significant bit planes. The more significant a bit is, the more structure is likely to be found in the corresponding bit plane. The structure is also similar from one bit plane to the next. This is illustrated in Figure 5.3. If visual features of the image are clearly recognisable in bit plane 7 and disappear completely in bit plane 8, it is reasonable to suspect LSB embedding. If one investigated bit planes 7 and 8 of the gorilla image in Figure 5.1, one would see that both are completely random. Another obvious reason to investigate more than one bit plane is that some steganographic algorithms embed in more than one bit plane.

The BSM features we will discuss in Section 5.3 are based on the observation that the structure in bit planes 7 and 8 is very similar. The features are designed to measure this similarity. Code Example 5.1 shows how Python creates the bitmap images shown in Figures 5.1–5.3.

5.2 Autocorrelation Features

Some of the artifacts we could observe visually by viewing the LSB plane as a two-tone image can be captured in the autocorrelation matrix, as suggested by Yadollahpour and Naimi (2009).

Code Example 5.1 Given an 8-bit image X in the form of a numpy array, return an 8-bit image representing the bth bit plane

---

```
def bitplane(X,b):
  b = 8 - bp
  return 255*((X>>b)%2)
```

The autocorrelation matrix A can be calculated for any 2-D signal S. Each element $A_{x,y}$ of A is the autocorrelation of S at a shift by some vector (x, y), that is

$$A_{x,y} = \sum_{i,j} S_{i,j} \cdot S_{i+x,j+y}.$$

In other words, each sample of S is multiplied by the sample x steps below and y steps to the right, and the result is summed over all samples.

Autocorrelation can be calculated either cyclically, by calculating the index $i + x$ modulo M and $j + x$ modulo N, in which case the summation is over the range $1 \leq i \leq M$ and $1 \leq j \leq N$. Alternatively, we can truncate and just limit the summation so that $1 \leq i, i + x \leq M$ and $1 \leq j, j + y \leq N$. Yadollahpour and Naimi (2009) did not specify which definition they had used. However, this is not likely to make a significant difference other than for very small images or very large step size (x, y). Code Example 5.2 shows how to calculate the autocorrelation features with truncation.

Autocorrelation is mainly interesting if the signal has zero mean. For a binary signal it is customary to substitute -1 for 0, to get a 2-D signal where each sample is ± 1. To study the LSB plane we calculate the autocorrelation matrix from $S = 2 \cdot (I \bmod 2) - 1$ where I is the original image; thus $S_{x,y} = \pm 1$. Observe that when $S_{i,j} = S_{i+x,j+y}$ we get $S_{i,j} \cdot S_{i+x,j+y} = 1$ and a positive contribution to $A_{x,y}$. Otherwise, we get $S_{i_j} \cdot S_{i+x,j+y} = -1$ and a negative contribution.

Example 5.2.1 *Consider an image where the LSBs are independent and uniformly random, and the autocorrelation $A_{1,0}$. Each black (resp. white) LSB is followed equally often by a white and a black LSB. Thus the four pairs $(-1, -1)$, $(-1, 1)$, $(1, -1)$ and $(1, 1)$ are equally common. Two give a product of -1 and two give $+1$, and the sum will be $A_{1,0} = 0$.*

Code Example 5.2 Function to calculate the AC feature for image im with shift idx

---

```
def acfeature(im,idx=(1,0)):
  X = 2*(X%2).astype(int) - 1
  (M,N) = X.shape
  (x,y) = idx
  return np.sum( X[x:,y:] * X[:M−x,:N−y] )
```

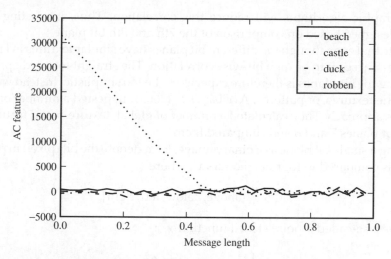

Figure 5.4 The AC statistic $A_{1,0}$ as a function of the embedded message length for a few images

Generally, an autocorrelation of 0 indicates that the samples are unrelated. Whether the autocorrelation is negative or positive, a high magnitude indicates some sort of structure. A positive correlation indicates significant monochrome areas, where pairs of pixel LSBs are equal and contribute positively. If the autocorrelation is negative, it means that LSBs alternate between 0 and 1, which is also a systematic pattern.

The experimental results shown in Figure 5.4 confirm what we have observed visually earlier in the chapter. Some images, such as the 'castle' image in Figure 5.3, have very high autocorrelation. Other images have approximately zero autocorrelation. For steganograms, with a sufficiently long message embedded, the autocorrelation is always zero, because the added noise erases any structure that might have been. In other words, autocorrelation is not an effective discriminant in the general case, but since it gives an indication for some images it may be a useful addition to a feature vector.

The most interesting elements are $A_{x,y}$ for small x and y, where high autocorrelation can be expected. Yadollahpour and Naimi (2009) proposed using $A_{1,1}$, $A_{1,2}$, $A_{1,3}$ and $A_{2,1}$, and they claimed experimental results showing that adding additional features from A did not have 'any considerable effect on classification accuracy'. We will denote Yadollahpour and Naimi's (2009) four-feature set by AC-4. It is surprising that horizontally and vertically adjacent pixels are not considered in AC-4. Therefore, we have also tested a larger feature vector comprising $A_{x,y}$ for $x + y \leq 4$ and $x, y \geq 0$. This gives ten features, and we denote it by AC-10.

5.3 Binary Similarity Measures

Figure 5.3 shows that there is a certain correlation between bit planes. Similar textures can be observed in different bit planes. By modifying only the LSB plane, many steganographic algorithms will reduce this correlation. Thus, interesting features may be designed based on comparison of the 7th and 8th bit plane.

We note that even though the different bit planes have similar textures in Figure 5.3, there does not seem to be high bit-wise correlation. The straight forward approach of correlating the bit planes is therefore expected to be too simplistic. Instead we need to capture the textures, or patterns. Avcibaş et al. (2005) suggested a number of features using this approach. They calculated a number of global, texture-based features from each of bit planes 7 and 8 and compared them.

Viewing a single bit plane as a binary image, let x_i denote the bit (pixel) in position i. Each bit is compared to four neighbours x_i^s, where

$$s \in S = \{above, below, left, right\}.$$

We define a so-called 5-point stencil function $\chi_{i,s}$:

$$\chi_{i,s} = \begin{cases} 1 & \text{if } x_i = 0 \text{ and } x_i^s = 0, \\ 2 & \text{if } x_i = 0 \text{ and } x_i^s = 1, \\ 3 & \text{if } x_i = 1 \text{ and } x_i^s = 0, \\ 4 & \text{if } x_i = 1 \text{ and } x_i^s = 1. \end{cases}$$

The local similarity measure for the pixel x_i is

$$\alpha_i^j = \sum_{s \in S} \delta(\chi_{i,s} = j), \text{ for } j = 1, 2, 3, 4.$$

Averaging over the bits in the image, we get four global variables $a = \bar{\alpha}^1, b = \bar{\alpha}^2, c = \bar{\alpha}^3$ and $d = \bar{\alpha}^4$. This could equivalently be described as making four one-dimensional passes, horizontal and vertical, forward and backward, and then adding the four one-dimensional measures $\chi_{i,s}$ together to get α_i^j. It follows that $b = c$, because a $(0, 1)$ pair in the forward pass becomes a $(1, 0)$ pair in the backward pass, and vice versa.

The features can be divided into three groups (i)–(iii), based on the different basic statistics and the way they are compared. The first group is an adaptation of classic one-dimensional similarity measures (Batagelj and Bren, 1995), and it consists of ten features. First, a 10-dimensional feature vector $(m_1, m_2, \ldots, m_{10})$ is calculated for each of the two bit planes, as follows:

$$m_1 = \frac{2(a + d)}{2(a + d) + b + c},$$

$$m_2 = \frac{a}{a + 2(b + c)},$$

$$m_3 = \frac{a}{b + c},$$

$$m_4 = \frac{a + d}{b + c},$$

$$m_5 = \frac{1}{4} \cdot \left(\frac{a}{a+b} + \frac{a}{a+c} + \frac{d}{b+d} + \frac{d}{c+d} \right),$$

$$m_6 = \frac{ad}{\sqrt{(a+b)(a+c)(b+d)(c+d)}},$$

$$m_7 = \sqrt{\frac{a}{a+b} \cdot \frac{a}{a+c}},$$

$$m_8 = \frac{b+c}{2a+b+c},$$

$$m_9 = \frac{bc}{(a+b+c+d)^2},$$

$$m_{10} = \frac{b+c}{4(a+b+c+d)}.$$

The ten group (i) features for the complete image are calculated as the vector difference of the features of the two bit planes:

$$m_i^{(i)} = m_i^7 - m_i^8,$$

where the superscripts 7, 8 indicate the bit plane.

Group (ii) is also calculated from the same basic statistics (a, b, c, d), but the two bit planes are compared using non-linear functions. Note that (a, b, c, d) are non-negative numbers whose sum is 4. This is seen as each statistic is the frequency of a type over the four neighbours of a pixel (averaged over all the pixels in the image). To be able to treat the statistics as probabilities, we divide by four and write

$$p_1^r = a/4, \quad p_2^r = b/4, \quad p_3^r = c/4, \quad p_4^r = d/4,$$

where $r = 7, 8$ indicates the bit plane. We compare p_i^7 and p_i^8 for each i before summing over i. The following four comparison functions are used to get four features:

$$f_1(z_7, z_8) = \min(z_7, z_8) \qquad \text{(minimum histogram difference)}$$

$$f_2(z_7, z_8) = |z_7 - z_8| \qquad \text{(absolute histogram difference)}$$

$$f_3(z_7, z_8) = -z_7 \cdot \log(z_8) \qquad \text{(mutual entropy)}$$

$$f_4(z_7, z_8) = -z_7 \cdot \log(z_7/z_8) \qquad \text{(Kullback–Leibler distance)}.$$

Thus, the four group (ii) features can be written as

$$m_j^{(ii)} = \sum_{i=1}^{4} f_j(p_i^7, p_i^8).$$

Remark 5.1 *It should be noted that groups (i) and (ii), totalling 14 features, have been calculated from only eight statistics (four statistics (a, b, c, d) from each bit plane). It is clear that these 14 features cannot contain any more information than the eight statistics whereon they are based, and thus it is reasonable to expect the feature vector to be highly redundant, and we shall indeed see that using the eight features instead of the 14 may give better performance experimentally.*

Group (iii), forming the last four features, is not calculated from the same basic features (a, b, c, d). Instead we use some kind of *texture units*, inspired by Ojala *et al.* (1996). Essentially a texture unit, for our purpose, is a possible value of the 3×3 neighbourhood around a given bit. There are $2^9 = 512$ possible texture units. We consider the texture unit of each pixel in the image, to calculate the histogram S_i^r where each texture unit is enumerated arbitrarily as $i = 0, \ldots 511$. As before, $r = 7, 8$ indicates the bit plane considered. Using the same comparison functions f_j as for group (ii), we can define the four group (iii) features as follows:

$$m_j^{(iii)} = \sum_{i=0}^{511} f_j(S_i^7, S_i^8).$$

Taking groups (i), (ii) and (iii) together, we get an 18-dimensional feature vector which we call BSM-18.

Note that the texture units used here are different from those used previously by Wang and He (1990) and Ojala *et al.* (1996). Originally, the texture units were defined for grey-scale images, recording whether each surrounding pixel is darker, lighter or equal to the central pixel. Avcibaş *et al.* (2005) modified the concept to be more suitable for binary images.

5.4 Evaluation and Comparison

We have compared the different feature sets in Table 5.1. As we saw with histogram analysis, short messages are almost undetectable, while embedding at full capacity is very easy to detect.

Table 5.1 Comparison of accuracies of feature vectors discussed in this chapter

Stego-system Message length	LSB	LSB±	LSB	LSB±	LSB	LSB±
	512B		40%		full length	
AC-4	49.6%	50.2%	62.1%	60.3%	82.1%	82.7%
AC-10	50.6%	51.1%	69.5%	70.7%	91.6%	92.0%
BSM-12	50.1%	51.1%	88.9%	74.4%	93.7%	92.9%
BSM-18	50.4%	49.3%	86.9%	72.7%	91.9%	92.2%
AC+BSM-22	50.5%	50.3%	88.8%	81.9%	93.8%	92.8%
ALE-10	50.1%	50.3%	79.1%	79.2%	95.7%	91.3%

We have formed the feature vector BSM-12 consisting of the four features in group (iii) from BSM-18, together with the preliminary statistics (a, b, c, d) from bit planes 7 and 8, as suggested in Remark 5.1. It is very interesting to see that this 12-feature set outperforms BSM-18 on our data, and the difference is statistically significant for 40% embedding and for LSB replacement at 100%.

The huge difference between LSB and LSB± at 40% for BSM is interesting. It is not surprising, as LSB replacement will never change bit plane 7, so that it can serve as a baseline measure when bit plane 8 is modified by embedding. For LSB± the embedding noise will also affect bit plane 7. It may be more surprising that we get good accuracy for both LSB and LSB± at 100%.

For the autocorrelation features, our experiments are not able to confirm the good performance reported by Yadollahpour and Naimi (2009). They did not give sufficient details to repeat their experiments. As they claimed to detect messages of length 0, it is natural to think that they had introduced difference in compression or post-processing which was picked up by the classifier instead of the embedding. We have also combined BSM-12 and AC-10 and observe that the resulting 22-D feature vector improves the accuracy for LSB± at 40% of capacity.

Comparing the new features to ALE-10, we see no consistent picture. In some cases, BSM-12 improves the accuracy, while ALE-10 prevails in others.

6

More Spatial Domain Features

We have looked at two fundamental concepts for the analysis of pixmap images, namely histograms and joint histograms, and individual bit planes. In this chapter we study two more, namely the difference matrices and image quality metrics (IQM). The image quality metrics were introduced by one of the pioneers in the use of machine learning for steganalysis, namely Avcibaş *et al.* (2003). The difference matrices are derived from joint histograms and have been used for several different feature vectors, including the recent subtractive pixel adjacency model (SPAM) features, which seem to be the current state of the art.

6.1 The Difference Matrix

The co-occurrence matrix, or 2-D histogram, from Chapter 4 contains a lot of information, with 2^{16} entries for 8-bit grey-scale images, and this makes it hard to analyse. Looking back at the scatter plots that we used to visualise the co-occurrence matrix, we note that most of the entries are zero and thus void of information. The interesting data are close to the diagonal, and the effect we are studying is the degree of spreading around the main diagonal; that is, the differences between neighbour pixels, rather than the joint distribution. This has led to the study of the difference matrix.

Definition 6.1.1 (Difference matrix) *The difference matrix* $D_{\delta_x,\delta_y} = D_{\delta_x,\delta_y}(I)$ *of an* $M \times N$ *matrix I is defined as*

$$D_{\delta_x,\delta_y}(x,y) = I[x + \delta_x, y + \delta_y] - I[x,y], \quad \text{for}$$

$$\max(1, 1 - \delta_x) \leq x \leq \min(M, M - \delta_x) \quad \text{and}$$

$$\max(1, 1 - \delta_y) \leq y \leq \min(N, N - \delta_y).$$

Machine Learning in Image Steganalysis, First Edition. Hans Georg Schaathun.
© 2012 John Wiley & Sons, Ltd. Published 2012 by John Wiley & Sons, Ltd.

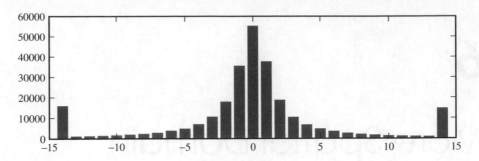

Figure 6.1   Histogram of the horizontal difference matrix $M_{0,1}$ of the 'beach' image

The difference matrix is simply the difference between the image matrix and a shifted version of the image matrix. The histogram of the difference matrix, as in the example in Figure 6.1, confirms the structure that we saw in the previous scatter plots. Most of the pixel pairs have zero difference, which means that they would be on the diagonal of the scatter plot. In fact, the histogram h_{δ_x,δ_y} of the difference matrix D_{δ_x,δ_y} is a projection of the corresponding co-occurrence matrix M_{δ_x,δ_y}, given as

$$h_{\delta_x,\delta_y}(k) = \sum_i M_{\delta_x,\delta_y}[i, i+k], \quad 1 - N \leq k \leq N - 1,$$

where N is the number of different grey-scale values.

In this section, we will first discuss the EM features of Chen *et al.* (2006b), based on the histogram and the HCF of the difference matrix. Then, we will discuss the SPAM features, which are based on a Markov model of the difference matrix.

6.1.1 The EM Features of Chen et al.

Chen *et al.* (2006b) proposed a feature set using both statistical moments and HCF moments of the histograms h_{δ_x,δ_y} of the difference matrices D_{δ_x,δ_y}.

Statistical Moments

Statistical moments are simple heuristics describing the shape of an empirical distribution or a PDF. The first couple of moments have simple intuitive interpretations. The first moment is the mean, which is also known as the centre of mass or as the expected value in the event of a probability distribution. The second moment is the variance, indicating the spread of the distribution. A very large variance indicates a distribution with long tails, whereas a low variance indicates a tall peak. The third one, ignoring scaling factors, is known as the skewness, telling us if and how the distribution is biased. Negative skewness indicates a long tail to the left, while positive skewness indicates a tail to the right. Symmetric distributions will have zero skewness.

Definition 6.1.2 *The k-th sample* moment *is defined as*

$$m_k = \frac{1}{n} \sum_{i=1}^{n} x_i^k,$$

where x_i are the sample observations for $i = 1, \ldots, n$.

Note that if $h(x)$ is the histogram of the observations x_i, then the definition above is equivalent to

$$m_k(h) = \frac{1}{N} \sum_{x=-255}^{+255} x^k h(x), \qquad (6.1)$$

where

$$N = \sum_{x=-255}^{+255} h(x)$$

is the total number of elements in the difference matrix D_δ. In statistics it is usually most useful to study the moments relative to the mean, and we define the sample central moments (also known as moments about the mean) as follows:

Definition 6.1.3 *The kth sample* central moment *is defined, for $k \geq 2$, as*

$$m_k = \frac{1}{n} \sum_{i=1}^{n} (x_i - \bar{x})^k,$$

where x_i are the sample observations for $i = 1, \ldots, n$, and $\bar{x}$ is the sample mean.

The EM-54 Features

The first 54 features of Chen *et al.* (2006b) are calculated from the image I itself, using nine different difference matrices D_{δ_x, δ_y}, for

$$(\delta_x, \delta_y) \in \Delta = \{(0,1), (0,2), (0,3), (1,1), (2,2), (3,3), (1,0), (2,0), (3,0)\}.$$

In other words, we consider pixel pairs at distance 1, 2 and 3, in three directions, horizontal, vertical and (main) diagonal. The histogram h_δ is calculated for $\delta \in \Delta$, and the HCF H_δ is calculated as the DFT of h_δ.

For each of the difference matrices D_δ, $\delta \in \Delta$, we use the moments of order 1, 2 and 3 of the histogram h_δ and the HCF H_δ to obtain the 6-D feature vector

$$g(D_\delta) = (m_1(h_\delta), m_2(h_\delta), m_3(h_\delta), m_1(H_\delta), m_2(H_\delta), m_3(H_\delta)).$$

The HCF moments $m_i(H_\delta)$ are calculated as before using (4.2) and the histogram moments using (6.1). It might have been more logical to use central moments instead of regular moments, but then D_δ is likely to have approximately zero mean so the difference may be negligible.

Concatenating $g(D_\delta)$ for $\delta \in \Delta$, we get a 54-D feature vector, denoted $f_{EM54}(I)$. We refer to the feature set as EM-54, where EM stands for the empirical matrix. It should be noted that the second moment of the difference matrices is equivalent to the gradient energy of the image, which was first introduced by Lin *et al.* (2004). A two-feature classifier by Lie and Lin (2005) used the sum $g = m_2(D_{1,0}) + m_2(D_{0,1})$ as the first feature.

Prediction Error Images and EM-108

Prediction error images is an idea from image compression which is much used in steganalysis. Pixels can be predicted by using information about neighbour pixels, giving the prediction image. In image compression, this means that instead of coding the pixel value, we only need to code the prediction error, i.e. the difference between the prediction and the actual value. Since most prediction errors are small, this can be stored more compactly.

Calculating the prediction error for every pixel gives a 2-D array which can be treated as an image, that is the prediction error image. In image steganalysis, prediction errors are interesting because they are more sensitive to embedding distortion than the image itself. Being independent of neighbour pixels, the embedding distortion cannot be predicted by the prediction algorithm. So, whereas the structure of the image itself disappears in the prediction error image, the characteristics of all the noise and distortion will be retained.

Two notable prediction techniques have been used in the steganalysis literature. The classic work of Lyu and Farid (2003) used linear regression to produce a prediction image. However, this predictor works in the wavelet domain, and will be considered in Chapter 7. Weinberger prediction, which we present below, applies in the spatial domain and has been used by many authors.

The lossless compression algorithm introduced by Weinberger *et al.* (1996) uses the following image prediction:

$$\hat{I}_{i,j} = \begin{cases} \max(I_{i+1,j}, I_{i,j+1}) & I_{i+1,j+1} \leq \min(I_{i+1,j}, I_{i,j+1}) \\ \min(I_{i+1,j}, I_{i,j+1}) & I_{i+1,j+1} \leq \max(I_{i+1,j}, I_{i,j+1}) \\ I_{i+1,j} + I_{i,j+1} - I_{i+1,j+1} & \text{otherwise.} \end{cases}$$

The prediction error image is defined by $E_{i,j} = I_{i,j} - \hat{I}_{i,j}$.

Note that the EM features $f_{EM54}(X)$ can be calculated from any matrix X, including the prediction error image E. This gives rise to the 108-D feature vector, EM-108, of Chen *et al.* (2006b). We use Weinberger prediction to calculate the prediction error image E and concatenate the 54 features for I and E, that is

$$f_{EM108}(I) = f_{EM54}(I) \| f_{EM54}(E),$$

where $\|$ denotes concatenation.

6.1.2 Markov Models and the SPAM Features

So far, we have discussed features based on the distribution of pixel pairs, either as the joint distribution or the distribution of differences. Natural images also have higher-order dependencies which may be exploited, and it is natural to look for features based on groups of more than two pixels.

The multi-dimensional histogram of joint distributions quickly becomes unmanageable, with 256^N entries for groups of N pixels in an 8-bit image. The difference matrix, in contrast is a good starting point. Where the pixel values are distributed more or less uniformly across 256 values when we consider a large collection of images, the difference matrix will have most of the values concentrated around zero. Thus, a histogram of the difference matrix requires much fewer bins to capture most of the information. This is the fundamental idea in the SPAM features of Pevný et al. (2009a) (see also Pevný et al. (2010b)). SPAM is short for *subtractive pixel adjacency model*. The *subtractive pixel adjacency matrix* is just a longer name for the difference matrix, and SPAM models this matrix as a Markov chain.

A Markov chain of order m is a sequence of stochastic variables $X_1, X_2, \ldots$ where the X_i are identically distributed, and X_i is independent of all X_j for all $j < i - m$. In other words, the distribution of an element X_i may depend on the m immediately preceding variables, but not on any older ones. The Markov chain is determined by specifying the transition probabilities $P(X_i | X_{i-m}, X_{i-m+1}, \ldots, X_{i-1})$.

In SPAM, the Markov chain is a sequence of adjacent differences in the difference matrix $D_{i,j}$, taken in the same direction as the difference. For instance, for $D_{0,1}$ we form the Markov chain horizontally left to right, and for $D_{1,-1}$ the chain goes diagonally from the upper right corner towards the lower left. A total of eight difference matrices $D_{i,j}$ are used, for

$$(i,j) \in \{(0,1), (0,-1), (1,0), (-1,0), (1,1), (1,-1), (-1,1), (-1,-1)\},$$

which includes all horizontal, vertical and diagonal differences of adjacent pixels. Note that each direction is taken both forwards and backwards.

The transition probabilities can be represented in the transition probability matrices $M_{i,j}$ and $M'_{i,j}$, as follows:

$$M_{i,j}[u,v] = \hat{P}(D_{i,j}(x+i, y+j) = u \mid D_{i,j}(x,y) = v),$$

$$M'_{i,j}[u,v,w] = \hat{P}(D_{i,j}(x+2i, y+2j) = u \mid D_{i,j}(x+i, y+j) = v; D_{i,j}(x,y) = w),$$

where $\hat{P}$ denotes the estimated probability based on empirical frequencies over all (x,y).

The difference matrix for an 8-bit image can have entries ranging from -255 to $+255$. Thus, the transition probability matrices are even larger than the multi-dimensional histograms of the image itself. However, since most entries will have low magnitude, we lose little by ignoring high-magnitude entries. We simply constrain the indices, as $-T \leq u,v,w \leq +T$, for some suitable threshold T. Thus, $M_{i,j}$ would have $(2T+1)^2$ entries, and $M'_{i,j}$ would have $(2T+1)^3$ entries.

In order to estimate the transition probabilities, we use Bayes' law and the joint histograms. Bayes' law gives us that

$$M_{i,j}[u,v] = \frac{\hat{P}(D_{i,j}(x+i,y+j) = u; D_{i,j}(x,y) = v)}{\hat{P}(D_{i,j}(x,y) = v)},$$

$$M'_{i,j}[u,v,w] = \frac{\hat{P}(D_{i,j}(x+2i,y+2j) = u; D_{i,j}(x+i,y+j) = v; D_{i,j}(x,y) = w)}{\hat{P}(D_{i,j}(x+i,y+j) = v; D_{i,j}(x,y) = w)}.$$

The individual, unconditional probabilities are estimated using the joint histograms, as follows:

$$M_{i,j}[u,v] = \frac{h_{i,j}(u,v)}{h_{i,j}(v)}, \tag{6.2}$$

$$M'_{i,j}[u,v] = \frac{h_{i,j}(u,v,w)}{h_{i,j}(v,w)}, \tag{6.3}$$

where $h_{i,j}$ are the one- and multi-dimensional histograms calculated from the difference matrix $D_{i,j}$. The multi-dimensional histograms are taken from the joint frequencies of pairs and triplets of pixels along the direction (i,j).

Pevný et al. (2009a) also assume that the statistics are approximately invariant under 90° rotation and mirroring. Therefore, there is no reason to maintain transition probabilities for eight directions. The four directions that are assumed to be approximately identical are averaged. Thus we get two matrices instead of eight, as follows:

$$F_1 = \frac{1}{4}[M_{0,1} + M_{0,-1} + M_{1,0} + M_{-1,0}],$$

$$F_2 = \frac{1}{4}[M_{1,1} + M_{1,-1} + M_{-1,1} + M_{-1,-1}].$$

Both F_1 and F_2 are $(2T+1) \times (2T+1)$ matrices. Similarly, we make F'_1 and F'_2 by averaging $M'_{i,j}$ in a similar way.

The final question to settle is the choice of threshold T. Pevný et al. (2009a) suggest $T = 4$ for the first-order transition probabilities F_1 and F_2, and $T = 3$ for the second-order probabilities F'_1 and F'_2. This gives a total of $2(2T+1)^2 = 162$ first-order SPAM features from F_1 and F_2, and $2(2T+1)^3 = 686$ second-order SPAM features from F'_1 and F'_2. We will refer to these feature sets as SPAM1-162 and SPAM2-686, respectively. Their union will be called SPAM-848.

The SPAM features have only been evaluated with respect to grey-scale images. It can be extended for colour images by calculating the features separately from each channel. Juxtaposing the three feature vectors gives about 2500 features, which may or may not be manageable. Pevný et al. (2010b) suggest training separate classifiers for each colour channel and then using fusion to combine the three. We will revisit this topic in Section 14.3.1.

6.1.3 Higher-order Differences

So far we have discussed features up to fourth order. The second-order SPAM features, using a second-order Markov model, consider the joint distribution of three adjacent entries in the difference matrix, where each entry is the difference of two pixels. Thus, four adjacent pixels are considered jointly in the calculation of the features.

The HUGO stego-system (Pevný et al., 2010a) was designed to preserve fourth-order statistics, and fourth-order features give poor classification. In their recent paper, Fridrich et al. (2011a) describe in some detail the process of developing an attack on HUGO. Their solution is to use features of order higher than four. This can be done in two different ways. We can use a higher-order probability model, such as a third-order Markov model to get fifth-order features, but this is computationally extremely expensive.

A less costly option is to go for higher-order differences. The second-order horizontal difference matrix $D = [d_{i,j}]$ for an image $X = [x_{i,j}]$ can be defined as

$$d_{i,j} = x_{i,j-1} - 2x_{i,j} + x_{i,j+1}.$$

Higher-order features can be calculated from second-order differences following the principles of SPAM, and third- and higher-order differences are possible as well. Fridrich et al. achieved the best results by combining a varied selection of higher-order features. Their paper is a good read for its frank report of the authors' own trial and error. The resulting feature vector was documented by Fridrich et al. (2011b).

6.1.4 Run-length Analysis

Run-length analysis was introduced by Dong and Tan (2008) and provides an alternative to Markov chains and joint histograms. The run length is well known from the compression algorithm used for fax transmissions, and it refers to the number of consecutive pixels with the same intensity, where the image is scanned as a 1-D sequence along some chosen direction (horizontal, vertical, major or minor diagonal). One run is defined as a set of consecutive pixels with the same value, and where the pixels immediately before and after the run have a different value. The run length is the number of pixels in the run. Counting all the runs gives rise to the run-length histogram h_{rl}, where $h_{rl}(i)$ is the number of runs of length i.

Without giving rise to equivalent features, the run-length histogram will be sensitive to embedding changes in the same way as difference matrices and joint histograms. Although relatively rare, long runs occur in natural images because of monochrome areas. Additive noise, such as from LSB embedding, would tend to break up such runs by introducing pixels which are one off the value of the original run.

Obviously, the run-length histogram cannot distinguish between a clean image with very strong texture, and a less textured image with an embedded message.

In either case there will be almost no long runs. This is probably the reason why Dong and Tan not only calculate the run-length histogram from the image itself, but also from a quantised image where the colour depth is reduced from 8 to 7 bits. While LSB replacement breaks up runs in the original image, no runs will be broken in the quantised image. A third run-length histogram is calculated from the difference image, where the runs are counted along the same direction as the differences.

The run-length features proposed by Dong and Tan (2008) use the histogram characteristic function. Following the approach of the HCF features, they take the first three HCF moments for each histogram. Using three different images (quantised, difference and original) and four directions (horizontal, vertical, minor and major diagonals), they get a 36-D feature vector. In their experiments it clearly outperforms HCF-78 and EM-108, but we have not seen a comparison with the SPAM features.

6.2 Image Quality Measures

Image quality measures are designed to measure the distortion between two versions of an image, thus they are functions $D(I, J)$ of an original image I and a distorted image J. Usually the distortion is due to compression or image coding, and the quality measure is used to evaluate the compression algorithm. Avcibaş et al. (2003) proposed a feature vector based on image quality measures they had previously studied in a different context (Avcibaş et al., 2002).

Obviously, the steganalyst has only one suspicious image I and no original to compare it against. Instead, Avcibaş et al. (2003) apply a Gaussian smoothing filter to the suspicious image I, to obtain a second image J. Then the features are calculated as image quality measures of I with respect to J. The filter is defined as follows:

$$H(m, n) = \kappa g(m, n) \quad -1 \leq m, n \leq +1,$$

$$g(m, n) = \frac{1}{2\pi\sigma^2} e^{-\frac{m^2 + n^2}{2\sigma^2}},$$

$$\kappa = \sqrt{\sum_m \sum_n |g(m, n)|^2},$$

where $g(m, n)$ is the 2-D Gaussian kernel and κ is the normalising constant ensuring unit norm. Avcibaş et al. (2003) suggest a mask size of 3×3. Based on experiments, they choose a standard deviation of $\sigma = 0.5$, but do not give any details.

Avcibaş et al. (2003) considered a total of 19 image quality measures which they evaluated by analysis of variance (ANOVA). Ten features survived the test, and we define these ten below for reference. All the features are defined in terms of colour images with K colour channels, but they remain well-defined for grey-scale images ($K = 1$). In only one case does the feature degenerate, giving $M_4 \equiv 0$ for $K = 1$.

Definition 6.2.1 *Let* $\|\mathbf{x}\|_\gamma$ *denote the γ-norm of $\mathbf{x}$, that is*

$$\|(x_1, x_2, \ldots, x_n)\|_\gamma = \left(\sum_{i=1}^{n} |x_i|^\gamma \right)^{1/\gamma}.$$

The γ-norm is also known as the Minkowsky norm of order γ. The 1-norm is also known as the taxicab norm and the 2-norm as the Euclidean norm.

The first two features are the 1-norm distance M_1 and the 2-norm distance M_2 between the two images, calculated for each colour channel and then averaged. Formally, we write it as follows:

$$M_\gamma = \frac{1}{K} \sum_{k=1}^{K} \left(\frac{1}{NM} \sum_{m=1}^{M} \sum_{n=1}^{N} |I_k(m, n) - J_k(m, n)|^\gamma \right)^{\frac{1}{\gamma}},$$

for $\gamma = 1, 2$.

The third feature is the Czekanowski distance, which is suitable for comparing vectors of strictly non-negative components. It is defined as follows:

$$M_3 = 1 - \frac{1}{MN} \sum_{m=1}^{M} \sum_{n=1}^{N} \frac{2 \sum_{k=1}^{K} \min(I_k(m, n), J_k(m, n))}{\sum_{k=1}^{K} (I_k(m, n) + J_k(m, n))}.$$

Avcibaş *et al.* (2003) did not specify how to deal with zero denominators in the fraction. However, they do point out that the fraction tends to one when $I_k(m, n) \to J_k(m, n)$, so it is reasonable to replace the entire fraction by 1 for any (n, m) where $I_k(m, n) = J_k(m, n) = 0$ for all k.

The fourth feature considers the angles between pixel vectors, i.e. RGB vectors. Since all colour intensities are non-negative, we are restricted to the first quadrant of the Cartesian space, and the maximum angle will be $\pi/2$. The angular correlation between two images is defined as follows:

$$M_4 = 1 - \frac{1}{MN} \sum_{m=1}^{M} \sum_{n=1}^{N} \frac{2}{\pi} \cos^{-1} \frac{\mathbf{I}(m, n) \cdot \mathbf{J}(m, n)}{\|\mathbf{I}(m, n)\| \cdot \|\mathbf{J}(m, n)\|},$$

where $\mathbf{I} \cdot \mathbf{J}$ denotes the inner product (scalar product) of two vectors, and $\| \cdot \|$ denotes the Euclidean norm.

Remark 6.1 *For grey-scale images M_4 is meaningless as the argument of $\cos^{-1}$ will always be 1.*

There are many variations of using the correlation to measure the similarity of two images. The next two features used are the image fidelity M_5 and the normalised cross-correlation M_6, defined as

$$M_5 = 1 - \left(\frac{1}{K} \sum_{k=1}^{K} \frac{\sum_{m=1}^{M} \sum_{n=1}^{N} (I_k(m,n) - J_k(m,n))^2}{\sum_{m=1}^{M} \sum_{n=1}^{N} I_k(m,n)^2} \right),$$

$$M_6 = \frac{1}{K} \sum_{k=1}^{K} \frac{\sum_{m=1}^{M} \sum_{n=1}^{N} I_k(m,n) J_k(m,n)}{\sum_{m=1}^{M} \sum_{n=1}^{N} I_k(m,n)^2}.$$

Avcibaş *et al.* (2003) used three spectral measures, i.e. features calculated from the discrete Fourier transform of the image. The DFT of the kth colour channel of the image I is defined as

$$\Gamma_k(u,v) = \sum_{m=1}^{M} \sum_{n=1}^{N} I_k(m,n) e^{-2\pi i m \frac{u}{M}} e^{-2\pi i n \frac{v}{N}},$$

for $k = 1, \dots, K$. The DFT of J is denoted by $\hat{\Gamma}_k(u,v)$, and it is calculated in the same way.

The first of the spectral measures considers the global transform image. It is the spectral magnitude distortion, given as

$$M_7 = \frac{1}{KMN} \sum_{k=1}^{k} \sum_{u=1}^{M} \sum_{v=1}^{N} \left| |\Gamma_k(u,v)| - |\hat{\Gamma}_k(u,v)| \right|^2.$$

The other two spectral measures are based on localised Fourier transforms. The image is divided into $b \times b$ blocks, where the inventors claim that $b = 32$ gives the most statistically significant results, and the DFT is calculated for each block.

Let $\Gamma_k^l(u,v)$ denote the transform of block l for $l = 1, \dots, L$, where L is the number of blocks. We write the transform block in magnitude-phase form as follows:

$$\Gamma_k^l(u,v) = |\Gamma_k^l(u,v)| e^{i\phi_k^l(u,v)},$$

where $\phi_k^l(u,v)$ is the phase of the complex number $\Gamma_l^k(u,v)$ and $|\Gamma_l^k(u,v)|$ is the absolute value. For each block l we calculate the average Euclidean distance of the magnitudes and the phases respectively, to give

$$J_M^l = \frac{1}{K} \sum_{k=1}^{K} \left\| \left| \Gamma_k^l(u,v) \right| - \left| \hat{\Gamma}_k^l(u,v) \right| \right\|,$$

$$J_\phi^l = \frac{1}{K} \sum_{k=1}^{K} \left\| \left| \phi_k^l(u,v) \right| - \left| \hat{\phi}_k^l(u,v) \right| \right\|.$$

Avcibaş *et al.* (2003) used the median of J_ϕ^l as their eighth feature:

$$M_8 = \operatorname*{median}_{l=1,\dots,L} J_\phi^l,$$

but apparently a similar median for J_M^l is found to be a very poor classifier. However, they do calculate a ninth feature based on a weighted average of J_M^l and J_ϕ^l as follows:

$$J^l = \lambda J_M^l + (1 - \lambda)J_\phi^l,$$

$$M_9 = \operatorname*{median}_{l=1,\dots,L} J^l.$$

The weighting factor λ is chosen to 'render the contributions of the magnitude and phase terms commensurate'. This gives some room for interpretation, but it is reasonable to assume that λ is calculated to give

$$\text{mean } \lambda J_M^l \approx \text{mean } (1 - \lambda)J_\phi^l,$$

where the means are taken over all blocks in all test set images.

Remark 6.2 *For small λ M_9 will be close to M_8, and for large λ it will be close to the median of J_M^l which was rejected as a feature. In either case M_9 is unlikely to give much information. The ANOVA test which was used to select features only considers individual features and ignores the effects, such as redundancy, of combining features.*

For the tenth feature, they use a human visual system (HVS) model proposed by Nill (1985). They assume that the human visual system can be modelled as a band-pass filter. They use a masking function, defined as

$$H(\rho) = \begin{cases} 0.05 e^{\rho^{0.554}} & \rho < 7, \\ e^{9|\log_{10}\rho - \log_{10}9|^{2.3}} & \rho \ge 7, \end{cases}$$

where $\rho = (u^2 + v^2)^{1/2}$, and calculate the HVS model $U\{I_k(i,j)\}$ of an image $I_k(i,j)$ as follows:

- calculate the 2-D discrete cosine transform (DCT) $\Omega_k(u,v)$ of $I_k(i,j)$ for each colour channel k;
- let $\Omega_k'(u,v) = H(\sqrt{u^2 + v^2})\Omega_k(u,v)$;
- let $U\{I_k(i,j)\}$ be the inverse DCT of $\Omega_k'(u,v)$.

The *normalised mean square HVS error* is defined as

$$M_{10} = \frac{1}{K}\sum_{k=1}^{K} \frac{\|U\{I_k\} - U\{J_k\}\|_2^2}{\|U\{I_k\}\|_2^2}.$$

6.3 Colour Images

Colour images make a much more complex question in steganalysis than grey-scale images. The complexity is not necessarily related to difficulty; more prominently it is a matter of having more options, which is likely to make steganograms more detectable but the research for the optimal approach more long-winded.

We have seen examples of how different authors have dealt with colour images differently. The simplest approach, commonly used to argue that a grey-scale technique generalises to colour images without necessarily testing it, is to extract features independently from each colour channel. This is trivial to implement and can be expected to work about as well as it does for grey-scale images.

Multiple colour channels give extra options which should make steganalysis more effective. A classic example of colour image steganalysis is due to Fridrich and Long (2000). It is founded on the assumption that most RGB images will only use a limited number of colours, out of the 2^{24} possible ones. LSB matching, and to some extent LSB embedding, will create new colours because each colour (r, g, b) can change into 26 other colours (r', g', b'), called neighbours of (r, g, b), with $|r - r'| \leq 1$, $|g - g'| \leq 1$ and $|b - b'| \leq 1$. This idea was developed further by Westfeld (2003), whose detector was more extensively evaluated by Ker (2005a). Westfeld stated that a colour in a clean image has 'only 4 or 5 neighbours on average'. Although this is likely to depend heavily on the quality and image format, it is true for some classes of images. LSB matching, even with relatively low embedding rate, will tend to create many colours where all 26 neighbours are in use. Westfeld suggests a single feature given as the maximum number of neighbours for any colour (r, g, b). If a feature vector is sought, other features can be calculated from the histogram of the number of neighbours per colour.

We have seen several examples which rely on the joint distribution over multiple colour channels. We discussed this in detail in relation to the HAR3D-3 and HAR2D-6 feature vectors, but it also applies to some of the IQM features discussed above. It may also be possible to generalise other features to capture the joint distribution of the colour channels. Relatively little research has gone into the evaluation and comparison of steganalysers for colour images, and most of the work that we can find is relatively old and does not consider the most recent feature vectors.

6.4 Experiment and Comparison

Table 6.1 shows empirical accuracies for the new features in this chapter. We observe that the SPAM features are able to detect very short messages; even though the accuracy is poor, it is significantly better than a coin toss. SPAM also appears to be significantly better than other feature vectors for longer messages, but several other feature vectors do almost as well for these messages.

The CMA features are calculated similarly to the SPAM features, using just $h_{i,j}(u, v)$ (for CMA1) and $h_{i,j}(u, v, w)$ (for CMA2) instead of the transition probabilities $M_{i,j}$ and $M'_{i,j}$. We note that the empirical data indicate that the SPAM features do indeed give the best classification. This is contrary to previous tests reported by

Table 6.1 Comparison of accuracies of feature vectors discussed in this chapter

Stego-system	LSB	LSB±	LSB	LSB±	LSB	LSB±
Message length	512B		40%		full length	
EM-54	48.8%	49.7%	89.1%	92.0%	96.2%	97.3%
EM-108	52.9%	52.7%	95.6%	96.0%	98.3%	98.1%
SPAM1-162	51.7%	53.7%	97.5%	94.4%	99.4%	98.4%
SPAM2-686	58.4%	59.7%	98.4%	97.2%	99.6%	99.0%
SPAM-848	56.9%	60.1%	98.5%	97.2%	99.5%	99.0%
CMA1-162	51.7%	52.6%	96.4%	92.8%	99.2%	97.5%
CMA2-686	56.1%	56.4%	96.1%	95.1%	99.2%	98.0%
CMA-848	54.8%	57.2%	96.1%	95.3%	99.3%	97.9%
IQM	51.2%	50.5%	62.0%	60.3%	71.7%	73.2%
ALE-10	50.1%	50.3%	79.1%	79.2%	95.7%	91.3%
BSM-12	50.1%	51.1%	88.9%	74.4%	93.7%	92.9%

Pevný *et al.* (2010a), who found similar performance and preferred to use the joint histograms for the sake of simplicity.

The IQM do badly in our tests. In the original work, Avcibaş *et al.* evaluated the IQM features on colour images and reported accuracies of about 70–75% against JSteg, Steganos and Stools, but their sample is only 22 images total for testing and training.

7

The Wavelets Domain

Wavelet transforms have acquired great popularity in digital image processing over the last two decades, and they may be best known for their application in JPEG2000 compression. The first machine learning-based steganalysis technique to gain widespread popularity, due to Lyu and Farid (2003), calculates the features from a wavelet representation of the images.

We will start with a high-level and visual view on wavelets, and how they are useful for image analysis. In Section 7.2, we explain how wavelet transforms are calculated in theory and in Python. The remaining sections will present common features which can be extracted from the wavelet domain, including the pioneering feature set of Lyu and Farid (2003).

7.1 A Visual View

An example of a one-level wavelet transform is shown in Figure 7.1(b), using the simple and well-known Haar wavelet. Visually, we can see that the upper left component of the transform image is an accurate reproduction at lower resolution. This is called the low-pass (or sometimes low-frequency) component L. The other three components are called high-pass components and give high-frequency information. The three high-pass components H, V and D represent local change in different directions, namely horizontal change (upper right), vertical change (lower left) and diagonal change (lower right), respectively. In the Haar wavelet, the high-pass components simply give the difference between one pixel and its neighbour.

The visual view shows instantly how wavelets are useful for image compression and transmission. Firstly, the high-pass components contain little information. They are dominantly uniformly grey and can be heavily compressed at little cost. Furthermore, in image transmission, we can transmit the low-pass component first, enabling the receiver to display a good approximation quickly. This low-resolution display is then refined as more and more high-frequency information is received.

Machine Learning in Image Steganalysis, First Edition. Hans Georg Schaathun.
© 2012 John Wiley & Sons, Ltd. Published 2012 by John Wiley & Sons, Ltd.

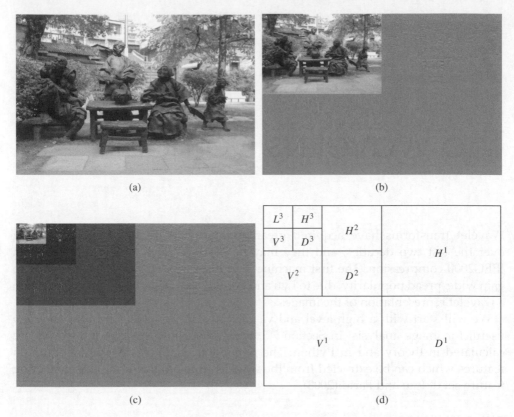

Figure 7.1 An image (a) and its two-level Haar wavelet transforms: (b) one-level transform; (c) three-level transform; (d) component labels. Photo by Welie and Hans Georg Schaathun © 2008

Each of the four blocks is obtained by applying one filter horizontally and one filter vertically. Each filter can be either a low-pass filter, meaning that it preserves low-frequency components while weakening high-frequency information or a high-pass filter, which conversely preserves the high-frequency components while weakening the low-frequency information. The low-pass component applies a low-pass filter in both dimensions. The three high-pass components are labelled according to the direction in which the high-pass filter is applied.

Higher-level wavelet transforms are obtained by applying the transform recursively on the low-pass component. See, for example, the three-level transform in Figure 7.1(c). We get one low-pass component L^3, and for each level $i = 1, 2, 3$ we get a horizontal component H^i, a vertical component V^i and a diagonal component D^i.

7.2 The Wavelet Domain

It is important to note that the wavelet transform is not one unique transform. There is a wide range of different wavelets, each defining a different transform. There are

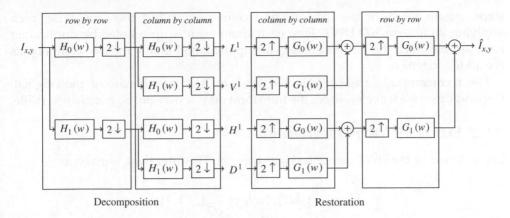

Figure 7.2 The fast wavelet transform (2-D)

also many different ways to explain and define wavelets and wavelet transforms, depending on the kind of understanding required.

Here we aim at a simple, procedural understanding appropriate for software implementation in a steganalysis system. A more complete introduction from an image-processing perspective can be found in Gonzalez and Woods (2008).

7.2.1 The Fast Wavelet Transform

The fast wavelet transform (FWT) is an efficient implementation of the discrete wavelet transform. It is also known as Mallat's herringbone algorithm (Mallat, 1989a, b). The 2-D FWT is presented schematically in Figure 7.2. The decomposition has two phases, one operating on rows and one on columns.

Each phase applies a 1-D FWT, which we describe as follows. Use a filter bank consisting of two filters: the low-pass filter H_0 and the high-pass filter H_1. The filters are viewed as signals $H_0(x)$ and $H_1(x)$, and they are applied to the input signal $I(x)$ by convolution. The signals (rows or columns) are filtered once with H_0 and once with H_1, to get a low- and a high-pass signal. Each of these filtered signals is downsampled (denoted 2 ↓). Thus, ignoring for now the special effects at the edges, the decomposition has the same bandwidth (number of samples) as the original signal, but the bandwidth is divided between a high- and a low-pass component.

After both the row and column passes, we have four components. They are sometimes labelled low–low (LL), low–high (LH), high–low (HL) and high–high (HH), indicating which filter was used in each phase. However, we will stick with the terminology already established, labelling the components by the direction of high-pass filtering, as L for low-pass (LL), V for vertical (LH), H for horizontal (HL) and D for diagonal (LL).

Decomposition is also known as analysis, and $\{H_0, H_1\}$ is called the *analysis filter bank*. In order to reconstruct the image we need inverse filters, that is a *synthesis filter bank*, denoted $\{G_0, G_1\}$ in the diagram. The reconstruction reverses the decomposition steps. Again we have two phases, first column-wise and then row-wise, each applying an inverse 1-D DWT. First each component is upsampled by duplicating each sample. Then the synthesis filters are applied, and finally the two components are added together.

The reconstructed image is denoted by $\hat{I}$, as it is an estimate of the original. Depending on the wavelets used, the transform may or may not be exactly reversible.

7.2.2 Example: The Haar Wavelet

Let us illustrate the DWT using the Haar wavelet. The filter bank is given as

$$H_0 = \frac{1}{\sqrt{2}}[1,1], \quad H_1 = \frac{1}{\sqrt{2}}[-1,1],$$

$$G_0 = \frac{1}{\sqrt{2}}[1,1], \quad G_1 = \frac{1}{\sqrt{2}}[1,-1],$$

where H_i are the analysis filters and G_i the synthesis filters. The scaling factor $1/\sqrt{2}$ is applied to allow us to use the same filters for analysis and synthesis. This is a common convention in the context of practical implementation, and we use it to be consistent with the pywt package in Python. In theoretical contexts the scaling factor is often omitted, and it would be common to see the Haar wavelet defined as $H_0 = [1,1]$ and $H_1 = [-1,1]$.

Consider first a 1-D sequence $S = (s_1, s_2, \ldots, s_n)$. Consider convolution with the analysis filters, and write

$$H_0 * S = Y = (y_1, y_2, \ldots, y_{n+1}), \tag{7.1}$$

$$H_1 * S = X = (x_1, x_2, \ldots, x_{n+1}). \tag{7.2}$$

It is easy to see that $y_i = (s_{i-1} + s_i)/\sqrt{2}$, where $s_0 = s_{n+1} = 0$. Similarly $x_i = (s_{i-1} - s_i)/\sqrt{2}$. Essentially, that is, ignoring the scaling factor of $\sqrt{2}$, the low-pass filter gives a sequence of sums of sample pairs, and the high-pass filter gives the differences.

Observe that if n is even, the elements involving s_0 and s_{n+1} disappear in the downsampling. This is the case for any filter of length 2. For longer filters, there are different ways of padding the image to avoid irregular effects near the borders. We'll have a quick look at this below.

Considering a 2-D image I, the filters will be applied first horizontally and then vertically. Denoting the four filtered images by $L_{x,y}$ (low-pass), $H_{x,y}$ (horizontal), $V_{x,y}$ (vertical) and $D_{x,y}$ (diagonal), we get the following expressions:

$$L_{x,y} = \frac{1}{2}(I_{2x,2y} + I_{2x+1,2y} + I_{2x,2y+1} + I_{2x+1,2y+1}),$$

$$H_{x,y} = \frac{1}{2}(I_{2x,2y} + I_{2x+1,2y} - I_{2x,2y+1} - I_{2x+1,2y+1}),$$

$$V_{x,y} = \frac{1}{2}(I_{2x,2y} - I_{2x+1,2y} + I_{2x,2y+1} - I_{2x+1,2y+1}),$$

$$D_{x,y} = \frac{1}{2}(I_{2x,2y} - I_{2x+1,2y} - I_{2x,2y+1} + I_{2x+1,2y+1}).$$

A pixel in the low-pass component is thus twice the average of 2×2 pixels in the source image. The horizontal component gives the difference between two horizontal neighbours, then averages this difference with that of the pair below it. Similarly, the vertical component takes the difference of a vertical pair and averages it with the horizontal neighbours. Finally, the diagonal component takes differences of differences.

In the H and V components we recognise the concept of difference matrices which we discussed in Section 6.1. The only difference is that the wavelet decomposition down samples the image, discarding every other difference. Yet, it is reasonable to expect that most of the information contained in features from the difference matrix can also be found in the horizontal or vertical components of a (one-level) wavelet decomposition with the Haar wavelet. Higher-order differences can also be construed as wavelet filters.

There are no detailed analysis or theoretical models explaining how steganography affects the wavelet coefficients, especially for the higher-level high-pass components. Studying H^1 and V^1 makes sense based on the analysis difference matrices in Section 6.1; however, the statistics will be different and a separate theoretical analysis and justification is called for. Similarly, the low-pass components can be related to using downsampling for calibration, which we will introduce in Section 9.3.

7.2.3 The Wavelet Transform in Python

The `pywt` package provides essential functions and objects to handle wavelet transforms in Python. The most important functions for image processing are `pywt.dwt2` and its inverse `pywt.idwt2`. For instance, to get the Haar wavelet decomposition of an image I, we can use

```
pywt.dwt2(I,"haar") ↦ ( L, (H, V, D) ),
```

where `L`, `H`, `V` and `D` are the four components L, H, V and D. The inverse call would be

```
pywt.idwt2((L,(H,V,D)),"haar") ↦ I.
```

The wavelet specification (`"haar"` in the example) can either be the name of a built-in wavelet as a string, or it can be a `Wavelet` object.

Custom wavelet objects can be generated by specifying the filter bank as an argument to the wavelet constructor. As an example, consider the code for creating the QMF-9 wavelet used for Farid's features in Code Example 7.1. We will discuss the calculation of the filter bank in Section 7.2.4. For the moment, we focus on the `Wavelet` constructor (in the last line). It is called `Wavelet(name,f)`, where `name` is a name for the wavelet and `f` is the filter bank, represented as a four-tuple with the

---

Code Example 7.1 Making a QMF-9 wavelet object in Python

```
lf = [ 0.02807382, −0.060944743, −0.073386624, 0.41472545, 0.7973934,
        0.41472545, −0.073386624, −0.060944743, 0.02807382, 0 ]
size = len(lf)
size2 = numpy.ceil(float(size)/2) − 1
hf = [ lf[i]*(−1)**(i−size2) for i in xrange(size−1,−1,−1) ]
f = ( lf, hf, list(reversed(lf)), list(reversed(hf)) )
qmf9 = pywt.Wavelet("qmf9",f)
```

Code Example 7.2 The Haar wavelet in Python

```
In [36]: w = pywt.Wavelet("haar")

In [37]: w.filter_bank
Out[37]:
([0.70710678118654757,  0.70710678118654757],
 [-0.70710678118654757,  0.70710678118654757],
 [0.70710678118654757,  0.70710678118654757],
 [0.70710678118654757,  -0.70710678118654757])
```

---

low-pass and high-pass decomposition and low-pass and high-pass reconstruction
filters in that order.

It is also possible to construct wavelet objects using the built-in wavelets. Then
we omit the filter bank argument and give just the name. The `filter_bank`
attribute gives access to the filter bank. See Code Example 7.2 for an example. The
`pywt.wavelist()` function returns a list of names of all the built-in wavelets.

Finally, multi-level decompositions can be made in one function call, using the
`pywt.wavedec2` function. The syntax is similar to that of `pywt.dwt2` but takes
an additional argument `level` specifying the number of decomposition levels. It
returns a tuple

$$(L, (H_i, V_i, D_i), (H_{i-1}, V_{i-1}, D_{i-1}), \ldots, (H_1, V_1, D_1)).$$

7.2.4 Other Wavelet Transforms

The study of wavelets is a complex one, with a wide range of different wavelets
having different properties. There is no consensus in the steganalysis community
about which wavelets to use, nor is it common to give reasons for the choice of
transform. The choice seems to be arbitrary, and no two groups appear to use the
same wavelet. We will just introduce a few wavelets and techniques which are
required for known steganalysis techniques.

We will see two other wavelets in use, besides the Haar wavelet, namely the well-known Daubechies wavelets and the less-known QMF-9 wavelet of Simoncelli and Adelson (1990). The Daubechies wavelets are readily available in Python, under the name dbX where X is the number of vanishing moments. The length of a Daubechies wavelet is twice the number of vanishing moments, so the 8-tap Daubechies wavelet we will use is available as pywt.Wavelet("db4").

Quadratic Mirror Filters

One wavelet, used for instance by Lyu and Farid (2003, 2006), is the QMF-9 wavelet of Simoncelli and Adelson (1990). This is an example of quadratic mirror filters (QMF), which is a class of filters distinguished by the property that

$$H_0(\omega) = G_0(-\omega) = F(\omega), \tag{7.3}$$

$$H_1(\omega) = G_1(-\omega) = e^{\omega i/2}F(-\omega + \pi), \tag{7.4}$$

for some function F, where i is the imaginary unit, i.e. $i^2 = -1$.

Simoncelli's QMF wavelets are not available in the pywt package, but Code Example 7.1 shows how to create a wavelet object for the QMF-9 wavelets used for Farid's features.

Padding at the Edges

When filters are applied using convolution, we implicitly assume infinite-length signals. With the images being finite, there will be strange effects at the border, where some of the filter elements should be multiplied by pixels outside the image boundary. Thus, we have to make some assumption and pad the image accordingly.

There are many ways to pad the image, the simplest and most common ones being the following:

$$(\ldots,0,0; \ x_1,x_2,\ldots,x_n; \ 0,0,\ldots) \quad \text{zero padding} \quad \text{"zpd"},$$

$$(\ldots,x_1,x_1; \ x_1,x_2,\ldots,x_n; \ x_n,x_n,\ldots) \quad \text{constant padding} \quad \text{"cpd"},$$

$$(\ldots,x_2,x_1; \ x_1,x_2,\ldots,x_n; \ x_n,x_{n-1},\ldots) \quad \text{symmetric padding} \quad \text{"sym"},$$

$$(\ldots,x_{n-1},x_n; \ x_1,x_2,\ldots,x_n; \ x_1,x_2,\ldots) \quad \text{periodic padding} \quad \text{"ppd"}.$$

The strings at the far right are the Python identifiers which can be passed to the wavelet decomposition functions as the mode argument. Somewhat more complicated paddings would extrapolate the signal using first derivatives, so-called smooth padding ("sp1") in Python.

The padding also means that the decomposed signal will have slightly greater bandwidth than the original image. That's necessary to ensure perfect reconstruction. Obviously, the padding probably makes little difference for our purposes, as we aggregate statistics over entire components and the padding only affects a small border region. In our experiments we have used the default in `pywt`, which is symmetric padding.

7.3 Farid's Features

The higher-order statistics of Lyu and Farid (2003, 2006) is one of the most cited techniques in the literature, and has inspired many other authors. A Matlab implementation of the feature extraction has been published on the Web (Farid, n.d). They reported tests using both FLD and SVM against JSteg, Outguess 0.1 and 0.2, EzStego and LSB in the spatial domain. Against JSteg and Outguess 0.1 they reached detection rates above 90% while keeping the false positives at 1%. It was less effective against the other three stego-systems.

7.3.1 The Image Statistics

Lyu and Farid make a wavelet decomposition using QMF-9. From each of the three high-pass components at levels 1, 2 and 3, they calculate four statistics. This gives 36 features for grey-scale images. For colour images, the image statistics are calculated from the wavelet compositions of each colour band separately, giving 36 features per colour band.

The statistics used are the mean, variance, skewness and kurtosis, which we define in terms of central moments around the mean. Recalling Definition 6.1.3, the kth sample central moment is given as

$$m_k = \frac{1}{n} \sum_{i=1}^{n} (x_i - \bar{x})^k,$$

where the x_i, in this case, are coefficients from one wavelet component. The sample variance is the second moment $m_2 = s^2$. The sample skewness and kurtosis are the third and fourth moments normalised by the variance, as follows:

$$g_1 = \frac{m_3}{m_2^{3/2}} \quad \text{(skewness)},$$

$$g_2 = \frac{m_4}{m_2^{2}} \quad \text{(kurtosis)}.$$

Note that the above definitions give biased estimators of the theoretical skewness and kurtosis. In the statistical literature, it is common to introduce a correction term to make unbiased estimators. However, for medium and large images the sample size is probably large enough to make the statistical bias negligible, and Farid (n.d.) used the unbiased estimators. Python code for the 36 features is shown in Code Example 7.3.

7.3.2 The Linear Predictor

In addition to the 36 image statistics per colour band introduced above, Lyu and Farid also consider a prediction error image. The idea is the same as we used for the

Code Example 7.3 Calculate the Farid features. For the definition of qmf9, see Code Example 7.1

```python
from scipy.stats import skew, kurtosis
from numpy import mean, var
def farid36(I):
  "Calculate_the_Farid_features_of_the_image_I."
  H = pywt.wavedec2( I, qmf9, level=4 )
  return farid36aux(H)
def farid36aux(H):
  R = []
  for h in H[2:]:
    R.extend( [ mean(A), var(A), skew(A), kurtosis(A) for A in h ] )
  return R
```

spatial domain in Section 6.1.1. The effect of embedding is very minor compared to the variability between images. By considering the prediction error image, we hope to capture just the noise, where the embedding distortion is larger relative to the signal. Lyu and Farid's (2003) prediction image is calculated using linear regression, as defined in any textbook on statistics, considering one wavelet component at a time. The definitions below are based on Farid's Matlab code (Farid, n.d.).

As an example, the vertical predictor for the red colour channel is given as follows:

$$|V_i^R(x,y)| = w_1|V_i^R(x-1,y)| + w_2|V_i^R(x+1,y)| + w_3|V_i^R(x,y-1)|$$
$$+ w_4|V_i^R(x,y+1)| + w_5|V_{i+1}^R\left(\frac{x}{2},\frac{y}{2}\right)| + w_6|D_i^R(x,y)| \tag{7.5}$$
$$+ w_7|D_{i+1}^R\left(\frac{x}{2},\frac{y}{2}\right)| + w_8|V_i^G(x,y)| + w_9|V_i^B(x,y)|,$$

where R, G and B refer to the red, green and blue colour channels respectively, and $|\cdot|$ denotes the element-wise absolute value. Assuming that the ith wavelet component is an $M \times N$ matrix, the ranges of indices x and y (7.5) are $x = 2,3,\ldots,M-1$ and $y = 2,3,\ldots,N-1$. Predictors for the green and blue channels are obtained by swapping R with G or B, respectively. For grey-scale images, the eighth and ninth terms are omitted.

The horizontal predictor is similar, substituting all the V-coefficients by corresponding H-coefficients. The diagonal predictor depends on both H- and V-coefficients, as follows:

$$|D_i^R(x,y)| = w_1|D_i^R(x-1,y)| + w_2|D_i^R(x+1,y)| + w_3|D_i^R(x,y-1)|$$
$$+ w_4|D_i^R(x,y+1)| + w_5|D_{i+1}^R\left(\frac{x}{2},\frac{y}{2}\right)| + w_6|H_i^R(x,y)| \tag{7.6}$$
$$+ w_7|V_i^R(x,y)| + w_8|D_i^G(x,y)| + w_9|D_i^B(x,y)|.$$

Again, blue and green predictors are obtained by swapping R with B or G, and a grey-scale predictor will use the first seven terms only.

The weights $\mathbf{w} = (w_1, w_2, \ldots, w_9)$ are calculated using the least-squares method. We will discuss the general technique in more detail in Chapter 12, so we will only briefly list the formulæ.

In order to calculate the predictor, we write (7.5) in matrix form:

$$\mathbf{v} = Q\mathbf{w},$$

where $\mathbf{v}$ is the elements of $|V_i^R|_{1<x<M,1<x<N}|$ strung out in vector form. Similarly, each column of Q contains the corresponding elements from one of the right-hand-side matrices. It can be shown that the following expression for the weights minimises the sum of squared errors:

$$\mathbf{w} = (Q^T Q)^{-1} Q^T \mathbf{v},$$

and the log prediction error, given as

$$\mathbf{p} = \log |\mathbf{v}| - \log |Q\mathbf{w}|,$$

where both the absolute value $| \cdot |$ and the logarithms are taken element-wise on the vectors. An implementation is shown in Code Example 7.4. The four features of each wavelet component are the mean, variance, skewness and kurtosis of the log prediction error vector $\mathbf{p}$, and they are calculated using the same technique as the first 36 features. The result is 72 features per colour channel, and we will refer to the feature vector as Farid-72.

7.3.3 Notes

Originally, Lyu and Farid (2003) only gave details for grey-scale images. The modified predictors for RGB images were first given explicitly by Lyu and Farid (2004). The most complete version of their work is found in Lyu and Farid (2006), including an additional set of features based on local angular harmonic decomposition (LAHD).

7.4 HCF in the Wavelet Domain

We discussed the use of the histogram characteristic function in the spatial domain in Chapter 4. The same principles can be applied in the wavelet domain. These ideas have been developed by Yun Q. Shi's group in New Jersey over a series of papers. The most recent and complete version appears to be in Shi *et al.* (2005a). We will first introduce the basic 39-feature set, HCF-39, of Xuan *et al.* (2005b).

Firstly, we decompose A using a three-level wavelet decomposition with the Haar wavelet. We will use all four components from each level i for $i = 1, 2, 3$, including the low-pass component L^i which is also decomposed further. Additionally, they include the original image (level 0), labelled $L^0 = A$. This gives 13 image components, from each of which we will extract three features.

Code Example 7.4 Farid's linear predictor. predError(hv,level) calculates the three log prediction error matrices **p**, corresponding to H, V and D from the level th component of the wavelet decomposition hv

```
def predError(hv,level):
    (H,V,D) = hv[−level]    # Level i components
    (H1,V1,D1) = hv[−(level+1)] # Level i+1 components
    (X,Y) = H.shape
    hx = [ i/2 for i in xrange(1,X−1) ]
    hy = [ i/2 for i in xrange(1,Y−1) ]
    D1 = D1 [hx , :] [: , hy]
    H1 = H1 [hx , :] [: , hy]
    V1 = V1 [hx , :] [: , hy]
    return (
        predAux( [ H[1:−1,1:−1], H[:−2,1:−1], H[1:−1,:−2],
            H1, D[1:−1,1:−1], D1, H[2:,1:−1], H[1:−1,2:] ] ),
        predAux( [ V[1:−1,1:−1], V[:−2,1:−1], V[1:−1,:−2],
            V1, D[1:−1,1:−1], D1, V[2:,1:−1], V[1:−1,2:] ] ),
        predAux( [ D[1:−1,1:−1], D[:−2,1:−1], D[1:−1,:−2], D1,
            H[1:−1,1:−1], V[1:−1,1:−1], D[2:,1:−1], D[1:−1,2:] ] ) )
def predAux(L):
    "Compute_the_linear_prediction_error_matrix_from_a_list_of_matrices."

    S = [ abs(A.flatten()[:,None]) for A in L ]
    v = abs(S[0])
    Q = np.matrix( np.hstack(S[1:]) )
    W = np.linalg.inv(np.transpose(Q)*Q) * np.transpose(Q) * v
    P = Q * W
    return log2(T) − log2(abs(P))
```

A schematic view of the entire procedure is shown in Figure 7.3, and Python code in Code Example 7.5. The HCF is calculated as discussed in Section 4.3.2, using each wavelet component in lieu of the pixmap matrix, as the DFT of the histogram. This gives us 13 HCFs. The HCF moments m_k are defined in Definition 4.3.5. The HCF-39 feature set is comprised of the kth moment m_k for $k = 1,2,3$, for each of the 13 components of the wavelet decomposition, for a total of 39 features. We write $f_{HCF39}(I)$ for the 39-D feature vector of the (pixmap) matrix I.

Figure 7.3 Diagrammatic overview of the HCF-39 feature extraction

Code Example 7.5 Function to create the HCF-39 feature vector

```
def hcf(C):
  "Calculate_the_moments_of_the_second-order_HCFs_of_C."
  C = C.flatten()
  nbins = max(C) - min(C) + 1
  nfreq = nbins / 2 + 1
  (h,b) = np.histogram( C, nbins )
  H0 = np.fft.fft(h)
  H = np.abs( H0[1:nfreq] )
  S = np.sum(H)
  return [ sum( [ (float(k)/nbins)**n*H[k] for k in xrange(nfreq) ] )/S
          for n in [1,2,3] ]
def hcf39(I,name="haar",level=3):
  """
  Calculate the HCF moments features of the image I.
  Optionally a wavelet name can be given as well.
  """
  w = getWavelet( name )
  R = hcf(I)
  for i in xrange(level):
    H = pywt.dwt2( I, w )
    for h in [H[0]] + list(H[1]):
      R.extend( hcf(h) )
    I = H[0]
  return R
```

Shi *et al.* (2005a) extended HCF-39 to a 78-feature set using a prediction error image. Given a (pixmap) image I, we calculate the prediction error $\hat{I}$ using Weinberger prediction as presented in Section 6.1.1. We get a 78-D vector, HCF-78, as

$$f_{HCF78}(I) = f_{HCF39}(I)\|f_{HCF39}(I - \hat{I}),\tag{7.7}$$

where $\|$ denotes concatenation.

7.4.1 Notes and Further Reading

Many papers have been written on HCF-78 and its predecessors. The most comprehensive is that by Xuan *et al.* (2005b) on HCF-39. The final version seems to be Shi *et al.* (2005a), which introduces the full 78-feature set. The theoretical justification for the moments of the characteristic function is discussed by Xuan *et al.* (2005a). An early version by Shi *et al.* (2005b) presented a smaller 18-feature set. Most of the papers use a Bayes classifier, but (Shi *et al.*, 2005a), finally concluded that neural networks give better results. Yun Q. Shi also has a web page (shi, 2009) giving an overview of the approach with examples.

We will return to the HCF features in Section 8.2.4, where we extend the feature set further by calculating HCF moments of the JPEG matrix, following Chen *et al.* (2006a).

7.5 Denoising and the WAM Features

A common task in signal and image processing is denoising, where the aim is to clean up a signal or an image and make it more usable by human observers through removing the noise. Once a denoised image is obtained, it is also possible to calculate the noise component, or *noise residual*, as the difference between the original and denoised images. Goljan *et al.* (2006) suggested an application in steganalysis, where the statistical features are calculated from the noise residual. Steganography by modification aims to make only imperceptible changes in the cover medium. If this objective is achieved, the denoised steganogram should be identical to the denoised cover image. Although practical stego-systems are unlikely to achieve the objective with perfection, it is still reasonable to expect the noise residual to be more sensitive to embedding.

The noise residual is closely related to prediction error images as used before (e.g. Section 6.1.1). Since the noise component cannot be predicted, all the noise will be included in the prediction error image, as it is in the noise residual. In a sense, the predictors used earlier can serve as crude denoising techniques without necessarily being designed for that purpose. Goljan *et al.* (2006), attributing the idea to an earlier two-page note by Holotyak *et al.* (2005), proposed using a technique designed specifically for denoising images.

There are many denoising algorithms available in the literature, and no hard and fast rule to select the best one for a given purpose. Both image quality and computational complexity may be important. Recently, Fridrich *et al.* (2011a) suggested Wiener filters, but we will only present the algorithm used in Goljan *et al.*'s original work on the WAM features.

7.5.1 The Denoising Algorithm

The denoising algorithm used for the WAM features is a variation of *locally adaptive window-based denoising* with maximum likelihood (LAW-ML), due to Mihçak *et al.* (1999a, b). For the sake of completeness, we will present both Goljan *et al.*'s version of LAW-ML and the original versions of Mihçak *et al.* We shall also see that the version originally used for the WAM features is not necessarily optimal. There is also a version using maximum a posteriori optimisation (LAW-MAP) (Mihçak *et al.*, 1999b), which is claimed to perform better for denoising, but this is left out for the sake of simplicity and space.

LAW denoising operates in the wavelet domain. The low-pass component, which is essentially a downsampled version of the image, is assumed to be sufficiently noise-free. Each of the high-pass components are considered separately, so let $X(x, y)$ denote the coefficients of one high-pass sub-band (horizontal, vertical or diagonal) of the unknown noise-free (cover) image, and let $Y(x,y)$ be the corresponding component of the received image with noise (steganogram). Our objective is to form an estimate $\hat{X}(x,y)$ of $X(x,y)$ based on $Y(x,y)$.

We model the high-pass coefficients of the (unknown) noise-free image as a series of independent Gaussian variables with zero mean and variance $\sigma^2(x,y)$, where (x,y) is the index within the wavelet component. The noise component is assumed to be additive white Gaussian noise of zero mean and variance σ_0^2. This gives us

$$Y(x,y) = X(x,y) + Z(x,y), \quad \text{where} \tag{7.8}$$

$$Z(x,y) \sim \mathcal{N}(0,\sigma_0^2),$$

$$X(x,y) \sim \mathcal{N}(0,\sigma^2(x,y)).$$

The variance $\sigma^2(x,y)$ is assumed to vary smoothly in x and y; we will refer to this as the smoothness assumption of the variance field. The distortion is assumed to have variance $\sigma_0^2 = 0.5$, which corresponds to (for instance) ± 1 embedding at 100% of capacity.

We choose our estimate $\hat{X}$ of X to minimise the mean square error (MSE) $E((\hat{X} - X)^2)$. It can be shown that an optimal estimate in this sense is given as

$$\hat{X}(x,y) = \frac{\sigma^2(x,y)}{\sigma^2(x,y) + \sigma_0^2} Y(x,y).$$

Unfortunately, the variance $\sigma^2(x,y)$ is unknown and must also be estimated.

The reliability of the estimate $\hat{X}$ will depend on the quality of the estimate $\hat{\sigma}^2(x,y)$ we use for $\sigma^2(x,y)$. The smoothness assumption of the variance field means that we can use the coefficients in a neighbourhood of (x,y) to estimate $\sigma^2(x,y)$. For instance, consider an $N \times N$ square neighbourhood around (x,y) for N odd. Noting that $E(Y) = 0$, a maximum likelihood estimator for the variance of $Y(x,y)$ is given as

$$s_N^2(x,y) = \sum_{i=x-\lfloor N/2 \rfloor}^{x+\lfloor N/2 \rfloor} \sum_{j=y-\lfloor N/2 \rfloor}^{y+\lfloor N/2 \rfloor} \frac{Y_{i,j}^2}{N^2}, \tag{7.9}$$

and with Y being the sum of two independent variables X and Z, we can estimate the variance of X as

$$\hat{\sigma}_N^2(x,y) = \max\left(0, s_N^2(x,y) - \sigma_0^2\right). \tag{7.10}$$

Python code for $N \times N$ LAW-ML is shown in Code Example 7.6.

Mihçak et al. (1999b) provides an argument that this estimate is an approximate maximum likelihood estimate of $\sigma^2(x,y)$. They also compared the performance of neighbourhood sizes $N = 3,5,7$. There was very little difference between 5×5 and 7×7 LAW-ML, whereas 3×3 LAW-ML was somewhat inferior. On the contrary, in the experiments at the end of this chapter, we shall see that 3×3 LAW-ML gives the best steganalyser on our data set.

Code Example 7.6 The function `lawmlN` takes a high-pass wavelet coefficient X and denoises it using $N \times N$ LAW-ML

```
def subm(X,N,i,j):
    "Return an NxN submatrix of X, centered at element (i,j)."
    N /= 2
    (m0,m1) = X.shape
    (i0,j0) = (max(0,i–N), max(0,j–N))
    (i1,j1) = (min(m0,i+N)+1, min(m1,j+N)+1)
    return X[i0:i1,j0:j1]

def lawmlN(X,N=3,sig0=0.5):
    "Return the denoised image using NxN LAW-ML."
    (m0,m1) = X.shape
    S = np.zeros((m0,m1))
    for x in xrange(m0):
        for y in xrange(m1):
            sub = subm(X,N,x,y)
            S[x,y] = max(0, np.var(sub) – sig0 )
    return X*S/(S+sig0)
```

7.5.2 Locally Adaptive LAW-ML

Image statistics may differ a lot between images and between different regions of the image. There is no reason to expect that one and the same neighbourhood size for LAW-ML will be optimal for every image or even throughout one image. Ideally, we would seek a more adaptive algorithms which uses the best possible estimator for $\sigma^2(x, y)$ for each (x, y). Both Mihçak *et al.* (1999a) in their original conference paper and Goljan *et al.* (2006) use adaptive estimators; and they do not use the same one.

Mihçak *et al.* (1999a) proposed to calculate $\hat{\sigma}_N^2$ for multiple neighbourhoods, $N \in S$ for some set S, and then choose the best estimate. Ideally, we would choose the estimate minimising the MSE $E(|\hat{X} - X|^2)$, but this problem is intractable. Instead, it is suggested to minimise the mean square error of the standard deviation estimate; that is to use $s_N^2(x, y)$ as the estimate, where

$$N = \arg\min_N E\left(|\sigma(x,y) - \hat{\sigma}_N(x,y)|^2 \right).$$

It is straightforward to show that this is equivalent to

$$N = \arg\min_N \left[(\sigma(x,y) - E(\hat{\sigma}_N(x,y)))^2 + \text{Var}(\hat{\sigma}_N(x,y)) \right].$$

According to Mihçak *et al.*, $\text{Var}(\hat{\sigma}_N(x,y))$ is the dominant term and they use just that in the minimisation, so

$$\hat{\sigma}^2 = \hat{\sigma}_N^2, \quad \text{where} \quad N = \arg\min_N \text{Var}(\hat{\sigma}_N(x,y)). \tag{7.11}$$

In order to estimate the variance of $\hat{\sigma}_N(x, y)$ they use the so-called bootstrap method. This is relatively simple, and works for small sample sizes; a detailed introduction can be found in Efron and Tibshirani (1993). The bootstrap method can be used to estimate the standard error (or variance) of any estimator $\hat{\theta}$, based only on the sample data used to calculate $\hat{\theta}$.

In our case we have N^2 pixel values from an $N \times N$ neighbourhood. Let $\mathbf{x} = (x_1, x_2, \ldots, x_{N^2})$ denote this sample of pixel values. This is used to calculate the estimate $\hat{\sigma}_N^2(x, y)$ as a function of $\mathbf{x}$ in (7.10). A bootstrap estimate of the variance $\mathrm{Var}(\hat{\sigma}_N^2(x, y))$ is calculated with the following algorithm.

1. Draw B samples $\mathbf{x}_i^*$ of N^2 elements each from $\mathbf{x}$ *with replacement*.
2. For each sample $\mathbf{x}_i^*$ calculate the estimator $\hat{\sigma}_N$, using $\mathbf{x}_i^*$ in lieu of the original sample $\mathbf{x}$ in (7.10). Let y_i denote the resulting value of $\hat{\sigma}_N$.
3. Given this sample of B instances y_i of the estimator $\hat{\sigma}_N$, we calculate the empirical variance of y_i ($i = 1, 2, \ldots, B$) to get

$$\widehat{\mathrm{Var}}(\hat{\sigma}_N) = \frac{\sum_{i=1}^{B}(y_i - \bar{y})^2}{B - 1}.$$

Using this algorithm to solve the minimisation problem in (7.11), we get an optimal estimate $\hat{\sigma}^2$ for σ_2.

The above algorithm is computationally very costly. In our implementation of adaptive LAW-ML, the denoising of a 1008×760 image took 1-2 hours per wavelet component (one standard desktop-range CPU). Although potentially practical for individual images, such a running time makes it virtually impossible to evaluate the denoising algorithm with a decent sample size. In the final journal paper, Mihçak *et al.* (1999b) completely ignored the adaptive algorithm, and Goljan *et al.* (2006) used a much faster algorithm with no comment on the differences. This is probably due to the computational cost.

Goljan *et al.* (2006) used the neighbourhood size giving the smaller estimated variance, i.e.

$$\hat{\sigma}_{\mathrm{GFH}}^2(x, y) = \min_{N \in \mathcal{S}} \hat{\sigma}_N^2(x, y). \tag{7.12}$$

This is very fast to compute, but in our experiments it is inferior to using a constant 3×3 neighbourhood size.

7.5.3 Wavelet Absolute Moments

The features proposed by Goljan *et al.* (2006) employ similar ideas to Lyu and Farid, but apply them to the noise residual. They use the central moments of the absolute values of each high-pass component to give a measure of the distribution of noise.

Thus, to calculate the features, we use the following algorithm. Firstly, calculate the first-level wavelet decomposition with the 8-tap Daubechies wavelet, to get three high-pass components H, V and D. Then, do the following steps for each component $X \in \{H, V, D\}$.

Code Example 7.7 The wam9 function calculates the WAM features from a single wavelet component C and works as an auxiliary for the wamNxN function, which calculates WAM(N)-27 for a given pixmap matrix. It depends on the denoising functions from Code Example 7.6

```
def wam9(C,N=3,nmom=9):
    "Calculate 9 WAM features from the wavelet component C."
    M0 = C - lawmlN(C,N)
    R = np.mean(M0)
    M = M0.flatten().astype(float) - R
    return [R] + [ np.sum(M**k)/M.size for k in range(2,nmom+1) ]

def wamNxN(I,N=3,nmom=9):
    "Calculate the WAM(N)-27 from the pixmap array I."
    (L,H) = pywt.dwt2(I,"db4")
    return reduce( list.__add__, [ wam9(C,N,nmom) for C in H ] )
```

1. Calculate the denoised component X'.
2. Calculate the noise residual $R = X - X'$.
3. Calculate the pth absolute central moment of R for $p = 1, 2, \ldots, n_{\mathrm{mom}}$, using the formula

$$m_k = \frac{1}{n} \sum_{i=1}^{n} |r_i - \bar{r}|^k ,$$

where r_i are the coefficients of R.

The algorithm is shown in Python in Code Example 7.7.

Goljan *et al.* (2006) used $n_{\mathrm{mom}} = 9$ moments. For grey-scale images, this gives a total of $3n_{\mathrm{mom}} = 27$ features. Note that, contrary to Farid's features, we use the absolute value $|x_i - \bar{x}|$ and do not normalise the moments by dividing by the variance.

Table 7.1 Comparison of accuracies of feature vectors discussed in this chapter

Stego-system Message length	LSB 512B	LSB± 512B	LSB 40%	LSB± 40%	LSB full length	LSB± full length
Farid-72	50.1%	51.6%	88.5%	87.8%	91.4%	91.7%
HCF-39	50.0%	50.0%	90.7%	83.7%	97.3%	93.3%
HCF-78	50.3%	53.1%	93.5%	91.4%	97.6%	97.6%
WAM(3)-27	49.1%	52.0%	91.9%	93.4%	94.2%	96.5%
WAM(7)-27	49.3%	52.1%	84.1%	82.7%	85.0%	89.8%
WAM(B)-27	50.9%	50.5%	85.0%	85.0%	88.6%	92.2%
SPAM-848	56.9%	60.1%	98.5%	97.2%	99.5%	99.0%

Remark 7.1 *Goljan* et al. *(2006) proposed the WAM features also for colour images by calculating the 27 features independently for each colour channel, for a total of 81 features.*

7.6 Experiment and Comparison

This chapter's empirical comparison is shown in Table 7.1. None of them can match the SPAM features. However, the different wavelet features perform remarkably well considering that they are not extracted from the same domain used for embedding.

8

Steganalysis in the JPEG Domain

Much of the research in steganography and steganalysis has focused on JPEG images, due to the sheer popularity of the format. It has been in widespread use, known to expert and non-expert users alike, for more than 15 years. Even though other formats may claim similar popularity at one time or another, they have been more short-lived.

Conventional wisdom indicates that steganalysis is most effective when the features are calculated directly from the domain of embedding. This is the 'firm belief' of authors like Goljan *et al.* (2006), and Fridrich *et al.* (2011a) attribute the idea to Fridrich (2005). The effect of the embedding can also more easily be understood in the embedding domain, and thus it is also easier, in general, to analyse the effect of embedding on features extracted from the same domain.

Since steganographic embedding in the JPEG domain is so popular, a considerable number of JPEG-based feature sets have emerged as well. The most well-known one seems currently to be the 219-dimensional feature vector PEV-219 (also known as PEV-274) of Pevný and Fridrich (2007), and the slightly simpler variation NCPEV-219. These feature vectors are very interesting not only because of their good performance, but also because they combine a range of different techniques. In fact, most of the features we introduce in this chapter are used in NCPEV-219 and PEV-219, and we will highlight those as we go along. Experiments have shown negligible performance difference between NCPEV-219 and PEV-219, so we will focus on the simpler variant. PEV-219 uses calibration, which is the topic of the next chapter.

8.1 JPEG Compression

JPEG is first and foremost a lossy compression system, aiming to discard information which is not significant to the human visual system. This system is comprised of a

Machine Learning in Image Steganalysis, First Edition. Hans Georg Schaathun.
© 2012 John Wiley & Sons, Ltd. Published 2012 by John Wiley & Sons, Ltd.

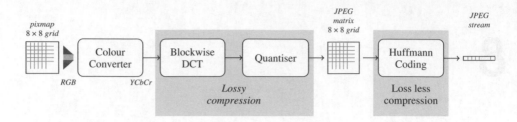

Figure 8.1 Schematic view of the JPEG compression process

series of transformations, as displayed in Figure 8.1. Each transformation has its own purpose, which we will explain below. Most stego-systems would operate between the lossy compression and the lossless coding, and consequently, this is also where we expect to find the most useful features.

8.1.1 The Compression

Several different colour models are supported by JPEG, including grey scale. In the case of colour images, it is customary, although not necessary, to use the YCbCr model. Where RGB has one component for each primary colour, red, green, and blue; YCbCr has one component Y for luminance or grey-scale information, and two colour (or *chrominance*) components Cb and Cr.

The advantage of the YCbCr model, compared to RGB, is that the human eye is more sensitive to changes in luminance (Y) than chrominance. Thus we may want to use a coarser compression on the Cb and Cr components. For instance, it is common to downsample, that is to reduce the resolution, of the Cb and Cr components while keeping the original resolution for the luminance component.

Each colour component (after any downsampling) is divided into 8×8 blocks. If necessary, the image is padded to get a size divisible by 8. Each 8×8 pixel block is mapped through a two-dimensional DCT transform, producing an 8×8 block of DCT coefficients.

The 2-D DCT transform can be viewed as writing the spatial 8×8 pixel matrix as a linear combination of 64 simple patterns, as shown in Figure 8.2. The $(0,0)$ coefficient (upper left-hand corner), also known as the DC coefficient, gives the average brightness in the block. The other coefficients are called AC coefficients, and indicate how the brightness changes across the block. Towards the lower right-hand corner, increasingly high-frequency changes are considered.

The low-frequency coefficients specify the coarse structure of the image, and changes here would be very perceptible. The high-frequency coefficient carries the fine detail, which we can afford to lose in compression. In the main compression step, or *quantisation*, each DCT coefficient block is divided, element-wise, by a quantisation matrix before the result is rounded to the nearest integer. High-frequency coefficients are divided by larger numbers, giving a more severe compression. We will refer to the integer matrix resulting from the quantisation step as the JPEG matrix, and to its

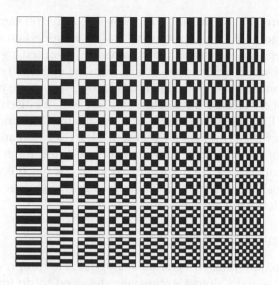

Figure 8.2 The DCT coefficients

elements as JPEG coefficients. The JPEG matrix is the data object we will work with throughout this chapter.

The quality or compression level can be controlled by the choice of quantisation matrix. Larger numbers give more compression and inferior image quality. The different colour channels may use different quantisation matrices, typically compressing chrominance more than luminance. All the quantisation matrices can be specified freely by the user, and they have to be stored in the JPEG file to allow decompression.

The most commonly used quantisation matrices are those used in the libjpeg library of IJG (IJG, 1998). This includes a set of base quantisation tables and a formula for scaling them using a quality factor $Q \in \{1, 2, \dots, 100\}$. If B_i is an entry in a base table, then the corresponding entry A_i in the scaled table is given as

$$A_i = \left\lfloor \frac{B_i \cdot S + 50}{100} \right\rfloor, \text{ where} \tag{8.1}$$

$$S = \begin{cases} 200 - 2Q & \text{for } Q \geq 50, \\ \frac{5000}{Q} & \text{for } Q \leq 50. \end{cases} \tag{8.2}$$

Thus, $Q = 50$ gives the base table unscaled, with higher numbers giving less and lower numbers more compression. It is common to give only the quality factor to specify how the compression is done, presumably assuming implicitly that the IJG tables are used. This is incomplete, though, both because other quantisation tables are possible, and because it says nothing about downsampling of the chromatic channels.

The final step is lossless compression, usually Huffmann coding. This step is an accurate one-to-one mapping, without any rounding error or lossy compression.

Hence the decompression algorithm will recover exactly the same JPEG matrix as the compression algorithm. The input to the lossless compression algorithm is a one-dimensional integer sequence, which is obtained by so-called *zig-zag scanning*. The JPEG matrix is read, one 8×8 block at a time, from the upper left corner zig-zagging along lines orthogonal on the main diagonal. As the stego-systems and feature sets we consider are not affected by the lossless step, we will not discuss it further.

Talking about a JPEG image, we will be referring to the JPEG matrix J, and we will let $J_{i,j}$ denote the coefficient in row i and column j, for $1 \leq i \leq M$ and $1 \leq j \leq N$ where $M \times N$ is the size of the matrix. The top row is row 1 and the left-hand column is column 1. Let $\mathcal{B}$ denote the set of 8×8 sub-blocks of J. If $\mathbf{b} \in \mathcal{B}$ is an 8×8 block, we let $b_{i,j}$ denote the coefficient in position (or frequency) (i,j), where the DC coefficient is $b_{0,0}$ in the upper left-hand corner.

8.1.2 Programming JPEG Steganography

It is easy to find programming APIs for most languages to compress and decompress JPEG images, but they are almost exclusively designed to do image processing in the spatial domain. Therefore, the image is normally decompressed upon loading, discarding the quantisation matrices, and a spatial domain representation is returned. Conversely, on saving an image, the input is a spatial domain image, which is compressed as an integral part of the serialisation into a JPEG file. It is hard to find an API to perform the intermediate steps individually, and get access to the JPEG matrix and other intermediate data structures.

Many popular implementations of stego-algorithms are implemented by tweaking the compression algorithm, simply adding the embedding steps between the lossy and the loss-less compression (cf. Figure 8.1). The cover image is loaded using a standard API, and in the event of JPEG input, it is automatically decompressed and the quantisation matrices are discarded. Consequently, the embedding algorithm has no information about the original quantisation matrices and will have to generate new ones. The resulting steganogram will most probably have been compressed twice with different quantisation matrices. This is known as *double compression*.

Double compression causes significant artifacts which tend to be even more detectable than the embedding itself. Software such as Andreas Westfeld's Java implementation of F5 and the classic JSteg implementation in C produce doubly compressed steganograms. When this software is used to generate steganograms for the training set of a machine learning algorithm, it is impossible to tell if it is the embedding or the double compression which is detected. Steganography software which does not double compress may be harder to detect.

It is important to note that double compression is an implementation artifact, and not a property of the algorithm. Neither JSteg nor F5 require any information from the spatial domain, and hence there is no reason to decompress a JPEG image. We should be looking for an API which can perform just the lossless step and give access to the JPEG and quantisation matrices. Although we are not aware of any production-quality software for the job, there are some alternatives which are good enough for research purposes. For example, Phil Sallee's JPEG Toolbox for MATLAB® (Sallee, 2009), and the pysteg library for Python, which is part of the companion software. Both are based on IJG's JPEG library and have very similar interfaces.

An example of accessing the JPEG data structure is shown in Code Example 8.1. The important feature is that we get direct access to both the coefficient matrices and the quantisation matrices. Usually, the luminance component uses the first coefficient matrix and the first quantisation matrix (index 0), but the JPEG file and the jpeg object also include a component info field which specifies which matrices are used for each colour channel. We shall see more examples of use throughout the chapter. We can note that this JPEG file is very high quality, as is seen by the small numbers in the quantisation matrix.

8.1.3 Embedding in JPEG

We briefly mentioned JPEG steganography in the discussion of LSB embedding in Chapter 2. The JPEG matrix, like the pixmap matrix, is an integer matrix and LSB embedding can be applied. The visual impact depends greatly on the coefficient value and the frequency. Changes to coefficients of low frequency and/or high magnitude would be most perceptible and it may be wise to avoid zeros and the DC coefficient.

The classic application of LSB embedding in JPEG is known as JSteg. To avoid the visual impact of modifying zero coefficients, it ignores 0 and 1, but it is otherwise identical to LSB replacement as discussed earlier. An implementation of JSteg, using the `jpeg` object from the previous section, is shown in Code Example 8.2. Note that it embeds in consecutive coefficients. Provos (2004) developed the technique further as software, which is often referenced, in two different versions. Outguess 0.1 follows JSteg but embeds in pseudo-random locations. We will use this algorithm in our evaluation at the end of the chapter. Outguess 0.2 introduced a new concept of statistics-aware embedding. Sacrificing capacity by not using all of the coefficients for embedding, Outguess 0.2 modifies unused coefficients in order to preserve statistical properties which are considered by known steganalysis attacks.

One of the most popular algorithms for JPEG steganography is F5, which was introduced by Westfeld (2001). Two new key techniques are introduced in F5. Firstly, the modulation is modified to preserve the symmetry of the histogram, resulting in an intermediate stego-system named F4. Secondly, Crandall's (1998) technique of matrix coding is employed to reduce the total number of modified coefficients.

A main problem of JSteg and Outguess 0.1 is that -1 coefficients may be changed to -2, while $+1$ coefficients are never modified. This results in an asymmetric histogram where clean images have symmetric histograms. In F4, a coefficient which needs to be modified is always modified by reducing the absolute value by one, thus pulling the coefficients towards 0, but maintaining symmetry. Zeros are ignored, and if a coefficient is changed to zero by the embedding algorithm, the corresponding message bit is re-embedded in the next available coefficient. The decoding algorithm

Code Example 8.1 Inspecting quantisation matrices and coefficient matrices in a JPEG image (high quality, colour) using pysteg. The output has been compressed to save space

```
In [1]: from pysteg.jpeg import jpeg
In [2]: J = jpeg( "duck.jpeg" )
[jpeg._ _init_ _] Image size 2592x3872
In [3]: J.quant_tables
Out[3]:
[array([[1, 1, 1, 1, 1, 1, 1, 2],
        [1, 1, 1, 1, 1, 2, 2, 2],
        [1, 1, 1, 1, 1, 2, 2, 2],
        [1, 1, 1, 1, 1, 2, 2, 2],
        [1, 1, 1, 2, 2, 3, 3, 2],
        [1, 1, 2, 2, 2, 3, 3, 3],
        [1, 2, 2, 2, 3, 3, 3, 3],
        [2, 3, 3, 3, 3, 3, 3, 3]], dtype=int32),
 array([[1, 1, 1, 1, 3, 3, 3, 3],
        [1, 1, 1, 2, 3, 3, 3, 3],
        [1, 1, 2, 3, 3, 3, 3, 3],
        [1, 2, 3, 3, 3, 3, 3, 3],
        [3, 3, 3, 3, 3, 3, 3, 3],
        [3, 3, 3, 3, 3, 3, 3, 3],
        [3, 3, 3, 3, 3, 3, 3, 3],
        [3, 3, 3, 3, 3, 3, 3, 3]], dtype=int32),
 None, None]
In [4]: J.coef_arrays
Out[5]:
[array([[610, -14, -1,..., 1, 1, 0],
        [  6, -4, -1,..., 0, 0, 0],
        [ -8, -1,  1,..., 0, 0, 0],
        ...,
        [ -3,  0,  0,..., 0, 0, 0],
        [  0,  0, -1,..., 0, 0, 0],
        [ -1,  0,  0,..., 0, 0, 0]], dtype=int32),
 array([[ 9, 1, 1,..., 0, 0, 0],
        [ -1, -1, -1,..., 0, 0, 0],
        [ 0, 0, 0,..., 0, 0, 0],
        ...,
        [ 0, 0, 0,..., 0, 0, 0]], dtype=int32),
 array([[-21, -3, 1,..., 0, 0, 0],
        [ 0, -1, 0,..., 0, 0, 0],
        [ 0, 0, 0,..., 0, 0, 0],
        ...,
        [ 0, 0, 0,..., 0, 0, 0]], dtype=int32)]
```

Code Example 8.2 Implementation of JSteg without double compression. The input J must be a jpeg object as demonstrated in Code Example 8.1 and msg should be a list or 1-D array of 0 and 1

```
def jsteg(J,msg):
    B = ( J < 0 ) | ( J > 1 )        # Mask out 0 and 1 coefficients
    B[::8,::8] = False               # Mask out DC coefficients
    S = J.coef_arrays[0][B]
    N = len(S)
    assert N >= len(msg), "Insufficient_capacity_in_cover_image"
    S[:N] -= S[:N] % 2               # Zero out LSB
    S[:N] += np.array( msg )         # Add message
    J.coef_arrays[0][B] = S
    return J
def dejsteg(J):
    B = ( J < 0 ) | ( J > 1 )        # Mask out 0 and 1 coefficients
    B[::8,::8] = False               # Mask out DC coefficients
    return J.coef_arrays[0][B] % 2
```

is modified to

$$
d_{F4}(b) = \begin{cases} b \bmod 2 & \text{for } b > 0, \\ b + 1 \bmod 2 & \text{for } b < 0, \\ \text{ignored} & \text{for } b = 0. \end{cases}
$$

The reason for using different decoding rules for negative and positive coefficients is that when both $+1$ and -1 are decoded as 1, there will be a lot of zero message bits that have to be re-embedded, while this never happens to message bits equal to one. With the decoding rule above, zeros and ones cause re-embedding equally often.

Matrix coding follows the ideas of coding for constrained memory as known from coding theory (e.g. Cohen *et al.* (1997)). JSteg uses one coefficient per message bit and modifies half a coefficient on average. F4 uses somewhat more, because of the re-embedding, but this is hard to quantify. Matrix coding increases the total number of coefficients required for the message, thus sacrificing capacity, but it reduces the number of changes, making the embedding harder to detect. Thus F5, which uses a simple matrix coding scheme on top of F4, will give steganograms that look like F4 with a shorter message embedded. Matrix coding has been developed further since, see for instance Kim *et al.* (2007).

In the experiments in this chapter, we will use our re-implementations of Outguess 0.1, which we label F1, and of F5. Contrary to most popular stego software available on the Internet, our implementations do not cause double compression.

8.2 Histogram Analysis

Much of the work on JPEG features has followed the lines of previous features from the spatial domain. This is natural, as in both cases we are looking at a 2-D signal, either the pixmap or the JPEG matrix, on which some embedding has made modifications. In most cases these changes appear as some kind of additive noise in the domain where the embedding operates. If the steganalyser operates in the same domain as the embedding, the analysis from Chapter 4 may apply.

A key difference between the spatial domain and JPEG is one of dimensionality. In the spatial domain, each coefficient (pixel) has a coordinate (x, y) in space, and we expect strong correlation between neighbour coefficients. In JPEG, we have sub-blocking, where the blocks have spatial coordinates (x, y), but where the individual coefficients have a frequency (u, v) within the block. Conceptually, the JPEG matrix is a 4-D array, with two spatial dimensions and two frequency dimensions. The blocks are located within the spatial dimensions, while the individual coefficients within a block are located in the frequency dimensions.

In the spatial domain, we expect heavy correlation between adjacent coefficients. Although similar dependencies may exist between adjacent JPEG coefficients too, this is overshadowed by the dependency of each coefficient on its frequency (u, v). Correlation is expected between neighbouring blocks, but as the distance between them is eight pixels, the correlation is less than what we can observe in the spatial domain. The 4-D structure of JPEG gives more options than the spatial domain, and we can consider both spatial relationships between blocks and relationships in the frequency domain.

8.2.1 The JPEG Histogram

First-order features can be calculated from the histogram of the JPEG coefficients. It is common to exclude the DC coefficients from the histogram, mainly because their distribution is very different from the AC coefficients, but possibly also because most stego-algorithms exclude them as well. The AC histograms shown in Figure 8.3 are typical. The distribution is symmetric around a very sharp peak at zero, and it can be modelled using a Laplace distribution (Lam and Goodman, 2000).

The histogram is greatly affected both by the quality factor (Figure 8.3) and the texture of the image (Figure 8.4). The impact of the quality factor is obvious. In order to compress the image more, we need to use coarser quantisation, and the result is to reduce the variance of the JPEG coefficients. In Figure 8.3 we see that all values except for zero are smaller for QF 50 than for QF 80. The significance of texture is equally obvious. Highly textured images like Baboon have a significant high-frequency component, which leads to large high-frequency coefficients in the JPEG matrix. Less textured images, like Lena, get a taller peak at zero.

Embeddings can affect the histogram in several ways (see Figure 8.5). The effect of JSteg is particularly obvious; the histogram is suddenly asymmetric. This is because $+1$ is excluded from embedding, while -1 may be changed to -2. We also have the

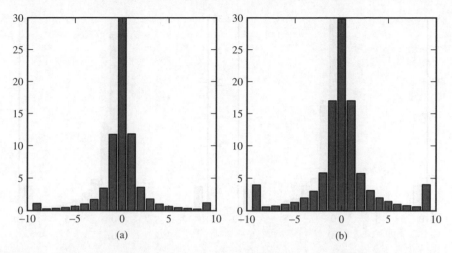

Figure 8.3 AC histograms for different quality factors: (a) QF 50; (b) QF 80. The units on the y-axis are in thousands. Note that the zero bar is out of range, and that the outermost bins are unbounded, representing the entire tail

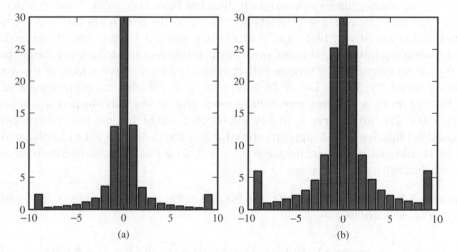

Figure 8.4 AC histograms for different JPEG images (QF 80): (a) Lena; (b) Baboon. The units on the y-axis are in thousands. Note that the zero bar is out of range

same effect as we saw when we studied the test in Section 2.3.4. The embedding will even out pairs of values in the histogram; e.g. -2 is increased and -1 is decreased, so that the two values come closer together.

The effect of F5 is more subtle. Westfeld (2001) designed F5 to mimic JPEG images at higher compression levels, and we can indeed see that the F5 steganogram has a heavier peak at zero than the natural image at the same quality factor. However, contrary

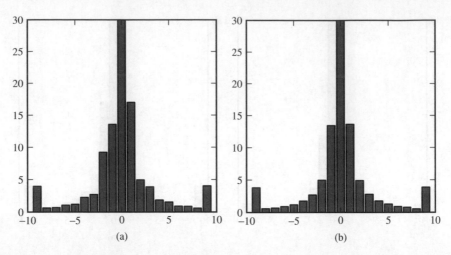

Figure 8.5 AC histograms for different embedding algorithms with a 4 Kb message: (a) F1; (b) F5. (Cover image as in Figure 8.3 (QF 80).) The units on the y-axis are in thousands. Note that the zero bar is out of range

to the QF 50 image, the F5 steganogram does not have noticeably fewer coefficients in the outermost bins. These artifacts are visible in the bar charts because we have embedded at a relatively high rate. With shorter message lengths, the visual artifacts will become negligible, but it is still possible to design features to identify the artifacts.

To give an example of a targeted discriminant, we will have a look at the asymmetry caused by JSteg. Let h be the histogram of the cover image, and h_q the histogram of a steganogram where JSteg embedding has flipped a fraction q of the LSBs. We can observe h_q in any intercepted image, but not the corresponding h. Based on this, we will design an estimator $\hat{q}$ of the distortion q. Evidently, a cover image would just be a special case with $q = 0$, so $\hat{q}$ would be a discriminant with $\hat{q} \approx 0$ indicating a natural image.

Theorem 8.2.1 (Symmetry attack) *The fraction q of modified coefficients in JSteg embedding can be estimated as*

$$\hat{q} = \frac{1}{2} - \frac{1}{2h_q(1)}\left[\sum_{i<0}\left(h_q(2i+1) - h_q(2i)\right) - \sum_{i>0}\left(h_q(2i+1) - h_q(2i)\right)\right],$$

where h_q is the histogram of the intercepted image.

Note that if the embedded message is a random, unbiased bit string then the message length can be estimated as $2\hat{q}$.

Proof. Consider one pair of values $(2i, 2i + 1)$ ($i \neq 0$). After embedding, we would have for the even value that

$$h_q(2i) = (1 - q)h(2i) + qh(2i + 1). \tag{8.3}$$

This can be seen because a fraction $(1 - q)$ of the $h(2i)$ coefficients remain unchanged, and a fraction q of the $h(2i + 1)$ coefficients have changed from $2i + 1$ to $2i$. Similarly, we have for the odd coefficients that

$$h_q(2i + 1) = qh(2i) + (1 - q)h(2i + 1). \tag{8.4}$$

The two equations, (8.3) and (8.4), can be rewritten as

$$h(2i) = \frac{1 - q}{1 - 2q}h_q(2i) - \frac{q}{1 - 2q}h_q(2i + 1), \tag{8.5}$$

$$h(2i + 1) = -\frac{q}{1 - 2q}h_q(2i) + \frac{1 - q}{1 - 2q}h_q(2i + 1). \tag{8.6}$$

There is one important exception, for $i = 0$. Since 0 and 1 are unmodified by JSteg, we have $h_q(0) = h(0)$ and $h_q(1) = h(1)$.

Because of the symmetry of the histogram, we have

$$\sum_{i>0} h(2i) \approx \sum_{i<0} h(2i) \quad \text{and}$$

$$\sum_{i<0} h(2i + 1) \approx h(1) + \sum_{i>0} h(2i + 1).$$

Adding the two equations, we have

$$h(1) + \sum_{i>0} h(2i + 1) + \sum_{i<0} h(2i) \approx \sum_{i<0} h(2i + 1) + \sum_{i>0} h(2i).$$

Substituting using (8.5) and (8.6) and rearranging, we get

$$h_q(1) \approx \sum_{i<0} \left[\frac{q}{1 - 2q}(h_q(2i + 1) - h_q(2i)) + \frac{1 - q}{1 - 2q}(h_q(2i + 1) - h_q(2i)) \right]$$

$$+ \sum_{i>0} \left[\frac{1 - q}{1 - 2q}(h_q(2i) - h_q(2i + 1)) + \frac{q}{1 - 2q}(h_q(2i) - h_q(2i + 1)) \right].$$

This can be simplified to get

$$(1 - 2q)h_q(1) \approx \sum_{i<0} (h_q(2i + 1) - h_q(2i)) - \sum_{i>0} (h_q(2i + 1) - h_q(2i)),$$

and the theorem follows.

□

8.2.2 First-order Features

Many different histograms may be useful for steganalysis. We have already mentioned the possibility of using either a complete histogram or an AC histogram. It can also be useful to consider per-frequency histograms, counting only coefficients of a particular frequency (u, v). The per-frequency histogram, also known as a local histogram, for a given frequency (u, v) can be defined as

$$h_{u,v}^d = \sum_{b \in \mathcal{B}} \delta(d = b_{u,v}),$$

where $\delta(E) = 1$ if the statement E is true, and $\delta(E) = 0$ otherwise. The global histogram and the AC histogram can be defined in terms of the local histograms as follows:

$$h_{AC}(d) = \sum_{(i,j) \neq (0,0)} h_{u,v}^d,$$

$$h_{glob}(d) = \sum_{i,j} h_{u,v}^d.$$

There are 64 local histograms, so it is hardly practical to use all of them.

The NCPEV-219 feature vector uses 11 elements from the global histogram and a total of 99 elements from nine low-frequency local histograms. The DC histogram is not used, but the next nine lowest frequencies are. From each of the histograms we use 11 central elements, as follows:

$$h_{glob}(d), \quad \text{for } d = -5, \dots, +5,$$

$$h_{u,v}^d, \quad \text{for } d = -5, \dots, +5, 1 \leq u + v \leq 3.$$

Calculation in Python is straightforward, as we can see in Code Example 8.3. Other variations of JPEG histograms have also been used, but sometimes they lead to redundant features.

The dual histogram is defined for each coefficient value d, as a 2-D matrix

$$\mathbf{g}_d = [h_{u,v}^d]_{u,v=0,1,2,\dots,7},$$

Code Example 8.3 Return the local histogram $h_{i,j}^d$ for $d = -T, \dots, +T$

```
def localHistogram(C,i,j,T):
    (h,b) = histogram( C[i::8,j::8].flatten(),
        np.array( [ float(i) - 0.5 for i in range(-T,T+2) ] ) )
    return h
```

---

where the local histogram, for comparison, can be written as a vector

$$\mathbf{h}_{u,v} = (h_{u,v}^d : d = \dots, -1, 0, +1, \dots),$$

for each frequency (u, v). Obviously, the same elements occur in the per-frequency histogram $\mathbf{h}$ and the dual histogram $\mathbf{g}$, just in different combinations. Using the histogram entries as features, the dual histogram clearly becomes redundant. NCPEV-219 includes 55 elements from the dual histogram, but they duplicate features from the local histograms. Therefore the feature vector was originally called PEV-274, even though the dimensionality is only 219.

The total number of non-zero coefficients is also sometimes used as a feature, but for a given image size the total number of non-zero coefficients holds exactly the same information as $h_{\mathrm{glob}}(0)$. Thus the total number of non-zero coefficients is redundant alongside the global histogram.

Normalisation of the histograms is an important issue which is rarely addressed explicitly in the literature. The standard approach seems to be to use the raw histograms, resulting in features heavily dependent on the image size. In the experiments reported in the literature, this may not be an issue as some fixed image size is normally used in the experiments. However, to build a steganalyser to work with images of arbitrary and variable size, one should aim to avoid features depending on properties which are irrelevant to the classification, such as image size. The problem is easily solved by normalising the histograms to use relative rather than absolute frequencies, i.e. we use $\mathbf{h}_{u,v}/\|\mathbf{h}_{u,v}\|_1$ and $\mathbf{g}_d/\|\mathbf{g}_d\|_1$ (Fridrich, 2005).

8.2.3 Second-order Features

The four-dimensional nature of the JPEG matrix gives several options for second-order features for JPEG steganalysis. Considering adjacent coefficients, should we mean adjacent frequencies or adjacent in space?

Different authors have considered different families of second-order features. In this section we will focus on second-order features derived from spatially adjacent coefficients. That is, we consider pairs of coefficients of the same frequency from adjacent blocks. In the framework from Chapter 4, we can calculate co-occurrence and difference matrices from the JPEG matrix J instead of the pixmap matrix, but to consider coefficients from adjacent blocks we need a step size of 8. Thus we consider the co-occurrence matrices $M_{0,8}$ and $M_{8,0}$, and difference matrices $D_{0,8}$ and $D_{8,0}$.

The NCPEV-219 features use features derived from the co-occurrence and difference matrices, but with a twist. The rows and columns of blocks wrap around, so that the last block in a row is considered adjacent to the first block on the next. This wrap-around has not been justified theoretically, but it may make the implementation somewhat simpler. For large images, the effect of the wrap-around is probably negligible. Furthermore, they average the horizontal and vertical co-occurrence matrices, leading to the following definition.

Definition 8.2.2 (JPEG co-occurrence matrix (Fridrich, 2005)) *The co-occurrence coefficients for a JPEG image J are given by*

$$C_{s,t} = \frac{1}{2\#\mathcal{B}} \sum_{u,v=0}^{7} \sum_{k=1}^{|\mathcal{B}|-1} \left(\delta(s = J_{u,v}^{R,k})\delta(t = J_{u,v}^{R,k+1}) + \delta(s = J_{u,v}^{C,k})\delta(t = J_{u,v}^{C,k+1}) \right),$$

where $J^{R,r}$ denotes the rth 8×8 block indexed in row scan order and $J^{C,r}$ denotes the rth 8×8 block in column scan order. A co-occurrence matrix can be formed as $\mathbf{C}_T(J) = [C_{s,t}]_{-T \le s,t \le T}$ for a suitable threshold T.

The PEV-219 feature set includes 25 co-occurrence features, namely $C_{s,t}$ for $-2 \le s,t \le 2$. Fridrich's definition aggregates the occurrences over all frequencies (u,v), and over vertical and horizontal pairs. Obviously, it would also be possible to consider separate co-occurrence matrices for each frequency and for each direction.

The JPEG co-occurrence matrix does not have as much structure as the co-occurrence matrix in the spatial domain. Instead of coefficient pairs clustering around a line, they tend to cluster around the origin with some rough symmetry. Adding noise to the JPEG matrix has the same effect as we saw in the spatial domain, in increasing the spread of the distribution. This can be seen in Figure 8.6. The most notable effect in the steganogram is a drift of scatter points down to the left. This is due to JSteg and Outguess modifying -1 coefficients but not $+1$ coefficients. Unfortunately, the spread observed depends more on the compression level than on the embedding. This can to some extent be alleviated by calibration, which we will investigate in Chapter 9.

The difference matrices are used to calculate the *variation* (Fridrich, 2005), which is simply the average of the entries of the two difference matrices considered. Again

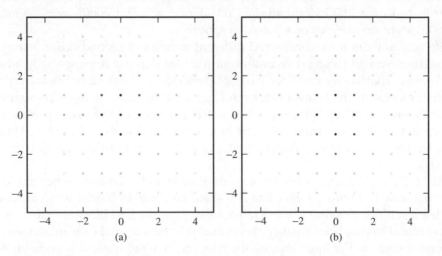

Figure 8.6 JPEG image with horizontal co-occurrence scatter diagram for frequency $(3,3)$: (a) cover; (b) Outguess 0.1

we have the little twist that the rows and columns wrap around. The result is the following definition which gives one feature for NCPEV-219.

Definition 8.2.3 (Variation (Fridrich, 2005)) *The variation $V(J)$ of an image J is defined as*

$$V = \frac{\sum_{u,v=1}^{8} \sum_{k=1}^{|\mathcal{B}|-1} |J_{u,v}^{R,k} - J_{u,v}^{R,k+1}| + \sum_{u,v=1}^{8} \sum_{k=1}^{|\mathcal{B}|-1} |J_{u,v}^{C,k} - J_{u,v}^{C,k+1}|}{2\#\mathcal{B}}.$$

We note that relatively few features are actually used, compared to the amount of information which is probably contained in the co-occurrence and difference matrices. Compared to 110 first-order features, we use only 26 second-order features. It could be an interesting question to see if this is optimal, or if further second-order features should be considered.

8.2.4 Histogram Characteristic Function

Naturally, the HCF can be used in the JPEG domain as it is in the spatial domain. Chen *et al.* (2006a) pursued this idea. Not only did they use the HCF of the JPEG matrix itself, they also used the HCF of wavelet decompositions of the JPEG matrix. Let $\mathfrak{J} = |J|$ be the element-wise absolute value of the JPEG matrix. We treat $\mathfrak{J}$ essentially like a pixmap matrix, and we can do wavelet decomposition and extract the HCF-39 feature vector $f_{\text{HCF39}}(\mathfrak{J})$ as defined in Section 7.4. This gives the first 39 HCF features from the JPEG domain.

We recall that another 39 features were calculated from a prediction error image. The Weinberger prediction algorithm (Section 6.1.1) was modified always to predict 0 when 0 is correct, i.e.

$$\hat{\mathfrak{J}}_{i,j} = \begin{cases} 0 & x = 0, \\ \max(\mathfrak{J}_{i+1,j}, \mathfrak{J}_{i,j+1}) & x \neq 0 \wedge \mathfrak{J}_{i+1,j+1} \leq \min(\mathfrak{J}_{i+1,j}, \mathfrak{J}_{i,j+1}), \\ \min(\mathfrak{J}_{i+1,j}, \mathfrak{J}_{i,j+1}) & x \neq 0 \wedge \mathfrak{J}_{i+1,j+1} \geq \max(\mathfrak{J}_{i+1,j}, \mathfrak{J}_{i,j+1}), \\ \mathfrak{J}_{i+1,j} + \mathfrak{J}_{i,j+1} - \mathfrak{J}_{i+1,j+1} & \text{otherwise.} \end{cases} \quad (8.7)$$

Based on this modified prediction image, we can calculate another 39 features as $f_{\text{HCF39}}(|\hat{\mathfrak{J}} - \mathfrak{J}|)$, and we have a total of 78 first-order features:

$$f_{\text{JHCF78}}(\mathfrak{J}) = f_{\text{HCF39}}(\mathfrak{J}) || f_{\text{HCF39}}(|\hat{\mathfrak{J}} - \mathfrak{J}|).$$

Chen *et al.* also added 234 features based on absolute marginal moments of a 2-D HCF. For each of the 13 wavelet components, three co-occurrence matrices $M_{0,1}$, $M_{1,0}$ and $M_{1,1}$ are considered. This gives a total of 39 co-occurrence matrices M, each of which gives rise to a 2-D HCF by taking the 2-D DFT of M. We need to generalise the concept of marginal moments for 2-D HCFs. This is analogous to the previous definition of the multi-dimensional centre of mass (Definition 4.4.1) and higher-order moments (Definition 4.3.5).

Definition 8.2.4 (Marginal moments) *Given a function g in d variables $\mathbf{z} = (z_1, z_2, \ldots, z_d)$, the marginal moments of order k are given as*

$$m_k(g(\mathbf{z})) = [c_1, c_2, \ldots, c_d], \quad where \tag{8.8}$$

$$c_i = \frac{\sum_{\mathbf{z}} z_i^n g(\mathbf{z})}{\sum_{\mathbf{z}} g(\mathbf{z})}. \tag{8.9}$$

The absolute marginal moments of g are $m_k(|g(\mathbf{z})|)$.

Each marginal moment gives two scalar features. Hence, the total number of features is 13 wavelet components times three co-occurrence matrices times three moments of order $n = 1, 2, 3$ times two features per moment; that is 234 features. We can summarise the algorithm as follows:

- Calculate three co-occurrence matrices $M_{0,1}$, $M_{1,0}$ and $M_{1,1}$ of X, defined as in Section 4.4.2.
- For each co-occurrence matrix $M_{i,j}$ take the 2-D DFT to get a 2-D HCF $H_{i,j}$.
- Calculate the absolute marginal moments $m_k(|H_{i,j}|)$ of order $k = 1, 2, 3$.

We label the resulting feature vector JHCF2D-234.

Combining HCF-78 from the wavelet domain with the 312 new HCF features from the JPEG domain, we get the HCF-390 feature set of Chen *et al.* (2006a):

$$f_{\text{HCF390}}(\mathfrak{I}) = f_{\text{HCF78}}(I) \| f_{\text{JHCF78}}(\mathfrak{I}) \| f_{\text{JHCF2D}}(\mathfrak{I}),$$

where $\mathfrak{I}$, and the pixmap matrix I, correspond to the same image $\mathfrak{I}$. Chen *et al.* obtained an accuracy comparable to that of FRI-23 (Fridrich, 2005) against Outguess and F5, and significantly better against MB1.

8.3 Blockiness

One disadvantage of embedding in the JPEG domain is the 8×8 block structure. The embedding will usually create independent distortion in each 8×8 block, and thereby break the continuity of the natural image at the block boundaries. The blockiness statistic (Fridrich *et al.*, 2002) was introduced to measure this effect. It was originally used for the non-learning *blockiness attack*, which was targeted on Outguess 0.2. Later, it was generalised to a family of blockiness features for use with machine learning.

In order to calculate the blockiness, the image is decompressed to give a pixmap matrix I. We consider all horizontal and vertical pixel pairs across any 8×8 block boundary, as indicated by the connecting edges in Figure 8.7. We define the vector $\mathbf{b}$, consisting of the pixel differences for each of these border pairs. The α-norm blockiness feature B_α is the normalised α-norm of $\mathbf{b}$. Formally, we can express it as follows.

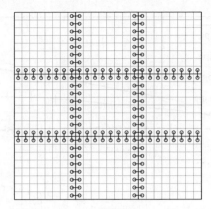

Figure 8.7 The pixels used to calculate blockiness

Definition 8.3.1 (Blockiness) *The α-norm blockiness feature is defined for any α as follows:*

$$B_\alpha(J) = \frac{\sum_{i=1}^{M_0} \sum_{j=1}^{N} |I_{8i,j} - I_{8i+1,j}|^\alpha + \sum_{i=1}^{M} \sum_{j=1}^{N_0} |I_{i,8j} - I_{i,8j+1}|^\alpha}{N \cdot M_0 + M \cdot N_0},$$

where $M_0 = M/8 - 1$, $N_0 = N/8 - 1$ and I is the pixmap matrix.

Code Example 8.4 shows how to calculate B_a in Python. The blockiness attack used just B_1, whereas PEV-219 uses B_1 and B_2.

Figure 8.8 illustrates how the blockiness statistic B_1 changes with the embedding length and different images. We note that F1 gives a very consistent increase in blockiness with increasing embedding length, whereas F5 has a less consistent effect. Highly textured images tend to have a higher blockiness than smoother images. For instance, the famous Baboon would have a much higher blockiness than any of our sample images. There is also a difference between different compression levels for the same image, although, in our experiments, this difference was less pronounced.

Code Example 8.4 Calculate the blockiness given a pixmap array X and the exponent α

```
def blockiness (X,alpha=1):
  (M,N) = np.shape(X)
  X = X.astype(float)
  A1 = np.abs(X[7:−8:8,:] − X[8:−7:8,:])**alpha
  A2 = np.abs(X[:,7:−8:8] − X[:,8:−7:8])**alpha
  B = np.sum(A1.flatten()) + np.sum(A2.flatten())
  return B / ( A1.size + A2.size )
```

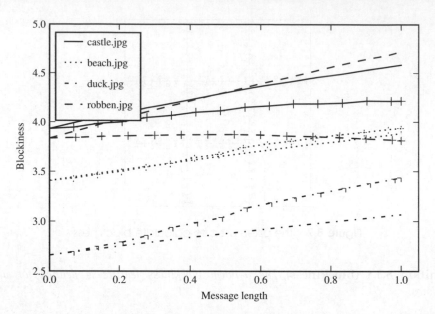

Figure 8.8 Comparison of blockiness at different embedding lengths and different images. The undecorated curves are for F1 and those decorated with + signs are for F5

8.4 Markov Model-based Features

The Markov model-based features were introduced by Shi *et al.* (2006), using the same framework that Pevný *et al.* (2009a) later used for the SPAM features that we saw in Section 6.1.2. As for the HCF features, we use the element-wise absolute values $\mathfrak{J} = |J|$ of the JPEG matrix J. From $\mathfrak{J}$, we define difference arrays as shown in Figure 8.9, which is analogous to the difference matrices in the spatial domain. This gives

$$M_h[i,j] = M_{0,1}[i,j] \ = \mathfrak{J}_{i,j} - \mathfrak{J}_{i+1,j},$$
$$M_v[i,j] = M_{1,0}[i,j] \ = \mathfrak{J}_{i,j} - \mathfrak{J}_{i,j+1},$$
$$M_d[i,j] = M_{1,1}[i,j] \ = \mathfrak{J}_{i,j} - \mathfrak{J}_{i+1,j+1},$$
$$M_m[i,j] = M_{-1,1}[i,j] = \mathfrak{J}_{i+1,j} - \mathfrak{J}_{i,j+1}.$$

The rest of the construction follows the design of the SPAM features.

For each of the difference arrays, we make a Markov model, considering a Markov chain along the direction of the difference, e.g. a horizontal Markov chain for the horizontal differences. In other words, we consider transition probability distributions

$$\left. \begin{array}{l} P(F_h(i+1,j) \mid F_h(i,j)), \\ P(F_v(i,j+1) \mid F_v(i,j)), \\ P(F_d(i+1,j+1) \mid F_d(i,j)), \\ P(F_m(i,j+1) \mid F_m(i+1,j)). \end{array} \right\} \tag{8.10}$$

Note that this Markov model completely ignores the frequencies of the coefficients.

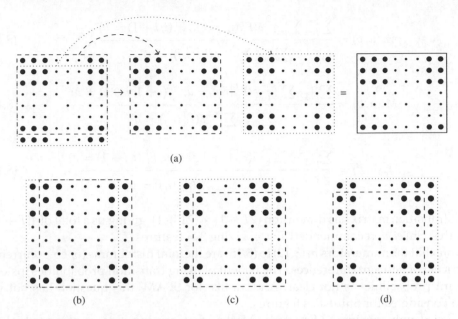

Figure 8.9 The difference arrays: (a) vertical differences; (b) horizontal; (c) major diagonal; (d) minor diagonal

Each row forms a separate Markov chain, so that the first item of the row is considered to be independent of the previous row. In contrast, the transition probabilities are assumed to be the same for all the row chains, and estimated based on all horizontal coefficient pairs. The same holds for columns and diagonals.

In principle, we want to estimate all the probabilities (8.10) empirically, and use them as features. However, the complexity would be high. The JPEG coefficients are unbounded, and we would get a great range of values in the difference arrays. In order to reduce the complexity of the Markov model, we define the T-capped difference arrays for $x = h, v, d, m$ as follows:

$$
F_x^T[i,j] = \begin{cases} T, & \text{if } F_x[i,j] > T, \\ -T, & \text{if } F_x[i,j] < -T, \\ F_x[i,j], & \text{otherwise,} \end{cases} \tag{8.11}
$$

and we define the Markov model over F_x^T instead of F_x.

The transition probabilities are estimated using the observed difference arrays and Bayes' theorem. Let $\delta(A) = 1$ if A is true and $\delta(A) = 0$ if A is false. The four transition probability matrices are defined as follows:

$$
M_h(u,v) = \frac{\sum_{i=1}^{S_i-2} \sum_{j=1}^{S_j} \delta(F_h^T(i,j) = u, F_h^T(i+1,j) = v)}{\sum_{i=1}^{S_i-1} \sum_{j=1}^{S_j} \delta(F_h^T(i,j) = u)}, \tag{8.12}
$$

$$M_v(u,v) = \frac{\sum_{i=1}^{S_i} \sum_{j=1}^{S_j-2} \delta(F_v^T(i,j) = u, F_v^T(i,j+1) = v)}{\sum_{i=1}^{S_i-1} \sum_{j=1}^{S_j} \delta(F_v^T(i,j) = u)}, \quad (8.13)$$

$$M_d(u,v) = \frac{\sum_{i=1}^{S_i-2} \sum_{j=1}^{S_j-2} \delta(F_d^T(i,j) = u, F_d^T(i+1,j+1) = v)}{\sum_{i=1}^{S_i-1} \sum_{j=1}^{S_j} \delta(F_d^T(i,j) = u)}, \quad (8.14)$$

$$M_m(u,v) = \frac{\sum_{i=1}^{S_i-2} \sum_{j=1}^{S_j+1} \delta(F_m^T(i+1,j) = u, F_m^T(i,j+1) = v)}{\sum_{i=1}^{S_i-1} \sum_{j=1}^{S_j+1} \delta(F_m^T(i,j) = u)}. \quad (8.15)$$

Each transition matrix has dimension $(2T + 1) \times (2T + 1)$, giving us a total of $4(T + 1)^2$ features. Shi *et al.* recommended $T = 4$, giving 324 features.

Note that the numerators in (8.12) to (8.15) are the joint histograms (or co-occurrence matrices) of adjacent differences. The benefit of using transition probabilities instead of joint probabilities is not clear. And, unlike for SPAM, this question has not yet been considered for Shi *et al.*'s features.

Shi *et al.* only considered first-order Markov features, but it does extend to higher-order models as well. Wahab *et al.* (2009) tested it with a second-order model. In other words, they considered transition probability distributions

$$P(F_h(i + 2, j)|F_h(i,j), F_h(i + 1, j)),$$

$$P(F_v(i, j + 2)|F_v(i,j), F_v(i, j + 1)),$$

$$P(F_d(i + 2, j + 2)|F_d(i,j), F_d(i + 1, j + 1)), \quad \text{and}$$

$$P(F_m(i, j + 2)|F_m(i + 1, j + 1), F_m(i + 2, j)).$$

Wahab *et al.* found that the second-order model has better accuracy in detecting double compression, but is slightly inferior in detecting F5 embedding in steganograms which have only been compressed once.

Shi *et al.*'s feature set has 342 features, which is a lot for closely related features. Designing PEV-219, Pevný and Fridrich chose to reduce the dimension by taking the average $\bar{M} = (M_h + M_v + M_d + M_m)$, which gives 81 features. As usual, there has been no research to see exactly how many Markov features it is reasonable to use.

8.5 Conditional Probabilities

The Shi–Chen–Chen Markov model totally ignores the block structure of the JPEG matrix, by assuming uniform transition probabilities across the JPEG matrix independently of the frequency of the coefficient. This assumption is clearly false. Most JPEG images will have almost only zeros in the lower right triangle of each 8×8

block, and the corresponding differences will mostly be zero as well. The highest magnitude differences will tend to be those crossing a block boundary or involving a DC coefficient.

When the transition probability matrix is estimated in the Shi–Chen–Chen framework, they aggregate coefficients of any frequency as if they had identical distribution. Theoretically, it makes more sense if each feature only considers coefficients at one location within the block. There are two approaches to this. Introducing the co-occurrence matrix $C_{s,t}$ above, we considered corresponding coefficients in neighbouring pairs of 8×8 blocks. Below we consider neighbouring coefficients within a block, taking one set of frequencies at time. The *conditional probability* feature vector (CP-27) was proposed by Briffa *et al.* (2009a). A key advantage of the CP features is that they give good classification with only 27 features, which makes the classifier much faster than the Shi–Chen–Chen classifier.

The CP-27 feature set considers three triplets (x_w, y_w, z_w) where w is h (horizontal), v (vertical) or d (diagonal) as shown in Figure 8.10. Nine features are extracted for each triplet. We consider the conditional probability distribution of the relationship between z and y, given the relationship between x and y. That is, we consider three prior events:

$$A_1^w = (x_w < y_w) \quad A_2^w = (x_w = y_w) \quad A_3^w = (x_w > y_w),$$

and three posterior events:

$$B_1^w = (y_w < z_w) \quad B_2^w = (y_w = z_w) \quad B_3^w = (y_w > z_w),$$

and we estimate the probabilities $P(B_i^w | A_j^w)$ for $i = 1, 2, 3$, $j = 1, 2, 3$ and $w = h, v, d$. The estimates are calculated using Bayes' law just like for the Markov model-based feature set. Hence, for each $w = h, v, d$, we count the number $h_{i,j}^w$ where events (A_i^w, B_j^w)

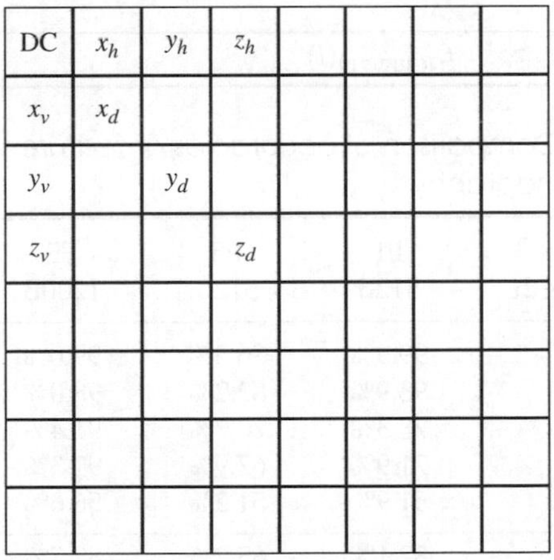

DC	x_h	y_h	z_h				
x_v	x_d						
y_v		y_d					
z_v			z_d				

Figure 8.10 The JPEG coefficients used for the CP feature set

occur together, and estimate the probability

$$\hat{P}(B_j^w \mid A_i^w) = \frac{h_{i,j}^w}{\sum_{j=1,2,3} h_{i,j}^w}.$$

Obviously, the CP feature set can be extended by adding additional coefficient triplets. In fact, the initial proposal of Briffa *et al.* (2009a) used six triplets (54 features), but the additional 27 features were taken from very high-frequency coefficients, and the 54-feature set was inferior to the 27-feature set in experiments.

8.6 Experiment and Comparison

We have covered a range of different features designed specifically to detect JPEG steganography. Since each group of features captures different information, an optimal classifier should almost certainly combine a range of different types of features. One example of this is the NCPEV-219 feature vector (Kodovský and Fridrich, 2009) defined in Table 8.1. NCPEV-219 has proved very good in experiments.

Table 8.1 Overview of the NCPEV features

Features	Notation		Number
Global histogram	$h_{\mathrm{glob}}(d)$	$-5 \le d \le 5$	11
Local AC histograms	$h_{u,v}^d$	$-5 \le d \le 5; 3 \le i+j \le 5$	11×9
Variation	V		1
Blockiness	B_i	$i = 1,2$	2
Co-occurrence	$C_{s,t}$	$-2 \le s,t \le 2$	25
NCPEV-138	$f_{\mathrm{NCPEV138}}(I)$		138
Markov features	$\mathbf{f_M}$		81
NCPEV-219	$f_{\mathrm{NCPEV219}}(I)$		219

Table 8.2 Comparison of accuracies of feature vectors for JPEG steganography

Stego-system Message length	F1 512B	F5 512B	F5 1200B	F5 4096B
NCPEV-219	99.9%	86.3%	99.0%	99.8%
Markov	98.9%	82.2%	98.0%	99.95%
JHCF	73.3%	70.7%	91.4%	98.6%
HCF-390	71.9%	67.9%	91.3%	98.0%
CP-27	51.9%	51.2%	56.6%	74.5%
SPAM	82.1%	65.2%	82.7%	97.5%
Farid	71.4%	53.4%	63.0%	87.0%

Test results for the feature vectors in this chapter are shown in Table 8.2. Several feature sets use a large number of closely related features, such as the Markov features and JHCF. These feature sets cannot meet the performance of NCPEV-219 and PEV-219, which are based on a very varied selection of features. Some recent works present new feature vectors which we have not had room to cover or test. For instance, Bhat *et al.* (2010) propose features based on the compression statistics of JPEG, both the statistics of the Huffmann code and the file size-to-image size ratio, and they claim to outperform PEV-219. Chen and Shi (2008) extended the Markov-based features to consider both interblock and intrablock correlations.

9

Calibration Techniques

One of the key challenges in steganalysis is that most features vary a lot within each class, and sometimes more than between classes. What if we could calibrate the features by estimating what the feature would have been for the cover image?

Several such calibration techniques have been proposed in the literature. We will discuss two of the most well-known ones, namely the JPEG calibration of Fridrich *et al.* (2002) and calibration by downsampling as introduced by Ker (2005b). Both of these techniques aim to estimate the features of the cover image. In Section 9.4, we will discuss a generalisation of calibration, looking beyond cover estimates.

9.1 Calibrated Features

We will start by considering calibration techniques aiming to estimate the features of the cover image, and introduce key terminology and notation.

We view a feature vector as a function $\mathbf{F} : \mathcal{X} \rightarrow \mathbb{R}^n$, where $\mathcal{X}$ is the image space (e.g. $\{0 \ldots 255\}^{M \times N}$ for 8-bit grey-scale images). A *reference transform* is any function cal : $\mathcal{X} \rightarrow \mathcal{X}$. Given an image I, the transformed image cal I is called the reference image.

If, for any cover image I_c and any corresponding steganogram I_s, we have

$$\mathbf{F}(\text{cal}\, I_s) \approx \mathbf{F}(I_c) \approx \mathbf{F}(\text{cal}\, I_c), \tag{9.1}$$

we say that cal is a *cover estimate* with respect to F. This clearly leads to a discriminant if additionally $\mathbf{F}(I_s) \not\approx \mathbf{F}(I_c)$. We could then simply compare $\mathbf{F}(I)$ and $\mathbf{F}(\text{cal}\, I)$ of an intercepted image I. If $\mathbf{F}(I) \approx \mathbf{F}(\text{cal}\, I)$ we can assume that the image is clean; otherwise it must be a steganogram.

The next question is how we can best quantify the relationship between $\mathbf{F}(I)$ and $\mathbf{F}(\text{cal}\, I)$ to get some sort of calibrated features. Often it is useful to take a scalar feature $F : \mathcal{X} \rightarrow \mathbb{R}$ and use it to construct a scalar calibrated feature $F_{\text{cal}} : \mathcal{X} \rightarrow \mathbb{R}$ as

Machine Learning in Image Steganalysis, First Edition. Hans Georg Schaathun.
© 2012 John Wiley & Sons, Ltd. Published 2012 by John Wiley & Sons, Ltd.

a function of $F(I)$ and $F(\text{cal } I)$. In this case, two obvious choices are the difference or the ratio:

$$F^D_{\text{cal}}(I) = F(I) - F(\text{cal } I),$$

$$F^R_{\text{cal}}(I) = \frac{F(I)}{F(\text{cal } I)},$$

where $F(I)$ is a scalar. Clearly, we would expect $F^D_{\text{cal}}(I) \approx 0$ and $F^R_{\text{cal}}(I) \approx 1$ for a natural image I, and some other value for a steganogram. Depending on the feature F, we may or may not know if the calibrated feature of a steganogram is likely to be smaller or greater than that of a natural image.

The definition of F^D_{cal} clearly extends to a vector function F; whereas F^R_{cal} would have to take the element-wise ratio for a vector function, i.e.

$$F^R_{\text{cal}}(I) = \left(\frac{f_i(I)}{f_i(\text{cal } I)} \bigg| i = 1, 2, \ldots, n \right), \quad \text{where}$$

$$F(I) = (f_i(I) \mid i = 1, 2, \ldots, n).$$

We will refer to $F^D_{\text{cal}}(I)$ as a *difference calibrated feature* and to $F^R_{\text{cal}}(I)$ as a *ratio calibrated feature*.

Figure 9.1 illustrates the relationship we are striving for in a 2-D feature space. The original feature vectors, $\mathbf{F}(I_c)$ and $\mathbf{F}(I_s)$, indicated by dashed arrows are relatively similar, both in angle and in magnitude. Difference calibration gives us feature vectors $\mathbf{F}^D_r(I_c)$ and $\mathbf{F}^D_r(I_s)$, indicated by solid arrows, that are 80–90° apart. There is a lot of slack in approximating $F(\text{cal } I) \approx F(I_c)$ for the purpose of illustration.

Calibration can in principle be applied to any feature F. However, a reference transform may be a cover estimate for one feature vector F, and not for another feature vector F'. Hence, one calibration technique cannot be blindly extended to new features, and has to be evaluated for each feature vector considered.

The ratio and difference have similar properties, but the different scales may very well make a serious difference in the classifier. We are not aware of any systematic

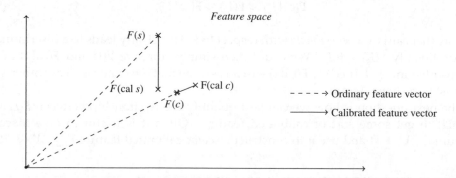

Figure 9.1 Suspicious image and (difference) calibrated image in feature space

comparison of the two in the literature, whether in general or in particular special cases. Fridrich (2005) used the difference of feature vectors, and Ker (2005b) used the ratio of scalar discriminants.

The calibrated features F_{cal}^R and F_{cal}^D are clearly good discriminants if the approximation in (9.1) is good and the underlying feature F has any discriminating power whatsoever. Therefore, it was quite surprising when Kodovský and Fridrich (2009) re-evaluated the features of Pevný and Fridrich (2007) and showed that calibration actually leads to inferior classification. The most plausible explanation is that the reference transform is not a good cover estimate for all images and for all the features considered.

The re-evaluation by Kodovský and Fridrich (2009) led to the third approach to feature calibration, namely the *Cartesian calibrated feature*. Instead of taking a scalar function of the original and calibrated features, we simply take both, as the Cartesian product $F_{cal}^C(I) = (F(I), F(\text{cal}\,I))$ for each scalar feature $F(I)$. The beauty of this approach is that no information is discarded, and the learning classifier can use all the information contained in $F(I)$ and $F(\text{cal}\,I)$ for training. Calibration then becomes an implicit part of learning. We will return to this in more detail in Section 9.2.2.

9.2 JPEG Calibration

The earliest and most well-known form of calibration is JPEG calibration. The idea is to shift the 8×8 grid by half a block. The embedding distortion, following the original 8×8 grid, is assumed to affect features calculated along the original grid only. This leads to the following algorithm.

Algorithm 9.2.1 *Given a JPEG image J, we can obtain the calibrated image* cal *J by the following steps:*

1. *Decompress the image J, to get a pixmap I.*
2. *Crop four pixels on each of the four sides of I.*
3. *Recompress I using the quantisation matrices from J, to get the calibrated image* cal *J.*

JPEG calibration was introduced for the blockiness attack, using 1-norm blockiness as the only feature. The technique was designed to be a cover estimate, and the original experiments confirmed that it is indeed a cover estimate with respect to blockiness. The reason why it works is that blockiness is designed to detect the discontinuities on the block boundaries caused by independent noise in each block. By shifting the grid, we get a compression domain which is independent of the embedding domain, and the discontinuities at the boundaries are expected to be weaker. This logic does not necessarily apply for other features.

9.2.1 The FRI-23 Feature Set

The idea of using JPEG calibration with learning classifiers was introduced by Fridrich (2005), creating a 23-dimensional feature vector which we will call FRI-23.

Most of the 23 features are created by taking the 1-norm $\|\mathbf{x}\|_1$ of an underlying, multi-dimensional, difference calibrated feature vector, as follows.

Definition 9.2.2 (Fridrich calibrated feature) *Let F be any feature extraction function operating on JPEG images. The Fridrich calibrated feature of F is defined as*

$$f_F(J) = \|F(J) - F(\text{cal } J)\|_1,$$

where $\|(x_1, \ldots, x_n)\|_1 = \sum_{i=1}^{n} \|x_i\|.$

Note that this definition applies to a scalar feature F as well, where the 1-norm reduces to the absolute value $|F(J) - F(\text{cal } J)|$. The features underlying FRI-23 are the same as for NCPEV-219, as we discussed in Chapter 8. Fridrich used 17 multi-dimensional feature vectors and three scalar features, giving rise to 20 features using Definition 9.2.2. She used a further three features not based on this definition. The 23 features are summarised in Table 9.1. The variation and blockiness features are simply the difference calibrated features from V, B_1 and B_2 from NCPEV-219.

For the histogram features, Definition 9.2.2 allows us to include the complete histogram. Taking just the norm keeps the total number of features down. The feature vectors used are:

- $(h_{\text{glob}}(d)|d = \ldots, -1, 0, +1, \ldots)$ (global histogram);
- $\mathbf{g}_d$ (dual histogram) for $d = -5, -4, \ldots, 0, \ldots, +5$; and
- $\mathbf{h}_{i,j}$ (per-frequency AC histograms) for $1 \leq i + j \leq 2$.

Each of these feature vectors gives one calibrated feature. Note that even though the same underlying features occur both in the local and the dual histogram, the resulting Fridrich calibrated features are distinct.

Table 9.1 Overview of the calibrated features used by Fridrich (2005) and Pevný and Fridrich (2007)

Base features	FRI-23		PEV-220	
	Features	No.	Features	No.
Global histogram	$\mathbf{h}_{\text{glob}}$	1	$h_{\text{glob}}(d)$	11
5 AC histograms	$\mathbf{h}_d$ $(-5 \leq d \leq +5)$	5	$\left.\vphantom{\begin{matrix}a\\b\end{matrix}}\right\} h_{i,j}^d$	11×9
11 Dual histograms	$\mathbf{h}_{i,j}$ $(1 \leq i+j \leq 2)$	11		
Variation	V	1	V	1
1-Norm and 2-norm blockiness	B_1, B_2	2	B_1, B_2	2
Co-occurrence	N_0, N_1, N_2	3	$\mathbf{C}_2$	25
Markov features	N/A	–	$\mathbf{f}_M$	81
JPEG non-zero	N/A	–	n_0	1
Total		23		220

The last couple of features are extracted from the co-occurrence matrix. These are special in that we use the signed difference, instead of the unsigned difference of the 1-norm in Definition 9.2.2.

Definition 9.2.3 (Co-occurrence features) *The* co-occurrence features N_i *are defined as follows:*

$$N_0(J) = C_{0,0}(J) - C_{0,0}(\text{cal } J),$$

$$N_1(J) = \sum_{\substack{i,j \\ |i|+|j|=1}} C_{i,j}(J) - C_{i,j}(\text{cal } J),$$

$$N_2(J) = \sum_{i=\pm 1} \sum_{j=\pm 1} C_{i,j}(J) - C_{i,j}(\text{cal } J).$$

According to Fridrich, the co-occurrence matrix tends to be symmetric around $(0, 0)$, giving a strong positive correlation between $C_{s,t}$ for $(s, t) \in \{(0, 1), (1, 0), (-1, 0), (0, -1)\}$ and for $(s, t) \in \{(1, 1), (1, -1), (-1, 1), (-1, -1)\}$. Thus, the elements which are added for N_1 and for N_2 above will tend to enforce each other and not cancel each other out, making this a good way to reduce dimensionality.

9.2.2 The Pevný Features and Cartesian Calibration

An important pioneer in promoting calibration techniques, FRI-23 seems to have too few features to be effective. Pevný and Fridrich (2007) used the difference-calibrated feature vector $F_{\text{cal}}^{\text{D}}(I)$ directly, instead of the Fridrich-calibrated features, where $F(I)$ is NCPEV-219.

The difference-calibrated features intuitively sound like a very good idea. However, in practice, they are not always as effective as they were supposed to be. Kodovský and Fridrich (2009) compared PEV-219 and NCPEV-219. Only for JP Hide and Seek did PEV-219 outperform NCPEV-219. For YASS, the calibrated features performed significantly worse. For the other four algorithms tested (nsF5, JSteg, Steghide and MME3), there was no significant difference.

The failure of difference-calibrated features led Kodovský and Fridrich to propose a Cartesian-calibrated feature vector, CCPEV-438, as the Cartesian product of NCPEV-219 calculated from the image J and NCPEV-219 calculated from cal J.

Table 9.2 shows some experimental results with different forms of calibration. The first test is based on Pevný's features, and it is not a very strong case for calibration of any kind. We compare the uncalibrated features (NCPEV-219), the difference-calibrated features (PEV-219), and the Cartesian features (PEV-438). We have also shown the features calculated only from cal J as CPEV-219. We immediately see that the original uncalibrated features are approximately even with Cartesian calibration and better than difference calibration.

A better case for calibration is found by considering other features, like the conditional probability features CP-27. Calibrated versions improve the accuracy

Table 9.2 Comparison of accuracies of feature vectors for JPEG steganography

Stego-system Message length	F1 512B	F5 512B	F5 1200B	F5 4096B
NCPEV-219	99.9%	86.3%	99.0%	99.8%
PEV-219	99.0%	81.4%	97.7%	99.9%
CCPEV-438	99.9%	86.7%	99.1%	99.9%
CPEV-219	54.3%	52.6%	63.3%	84.8%
FRI-23	95.5%	74.6%	93.3%	99.5%
CP-27	51.9%	51.2%	56.6%	74.5%
DCCP-27	60.9%	61.0%	75.4%	93.2%
CCCP-54	61.4%	61.9%	77.4%	95.2%

significantly in each of the cases tested. Another case for calibration was offered by Zhang and Zhang (2010), where the accuracy of the 243-D Markov feature vector was improved using Cartesian calibration.

When JPEG calibration does not improve the accuracy for NCPEV-219, the most plausible explanation is that it is not a good cover estimate. This is confirmed by the ability of CPEV-219 to discriminate between clean images and steganograms for long messages. This would have been impossible if $F(\text{cal}\,I_c) \approx F(\text{cal}\,I_s)$. Thus we have confirmed what we hinted earlier, that although JPEG calibration was designed as a cover estimate with respect to blockiness, there is no reason to assume that it will be a cover estimate with respect to other features.

JPEG calibration can also be used as a cover estimate with respect to the histogram (Fridrich et al., 2003b). It is not always very good, possibly because calibration itself introduce new artifacts, but it can be improved. A blurring filter applied to the decompressed image will even out the high-frequency noise caused by the original sub-blocking. Fridrich et al. (2003b) recommended applying the following blurring filter before recompression (between Steps 2 and 3 in Algorithm 9.2.1):

$$
M = \begin{bmatrix} 0 & e & 0 \\ e & 1-4e & e \\ 0 & e & 0 \end{bmatrix}.
$$

The resulting calibrated image had an AC histogram closely matching that of the original cover image, as desired.

Remark 9.1 *Also the run-length features (Section 6.1.4) can be seen as an application of Cartesian calibration. The motivation when we combined features from the original image and a quantised image is exactly the same as in the introduction of Cartesian calibration; the quantised image provides a baseline which is less affected by the embedding than the original image would be.*

9.3 Calibration by Downsampling

Downsampling is the action of reducing the resolution of a digital signal. In the simplest form a group of adjacent pixels is averaged to form one pixel in the downsampled image. Obviously, high-frequency information will be lost, while the low-frequency information will be preserved. Hence, one may assume that a downsampling of a steganogram I_s will be almost equal to the downsampling of the corresponding cover image I_c, as the high-frequency noise caused by embedding is lost. Potentially, this gives us a calibration technique, which has been explored by a number of authors.

Ker (2005b) is the pioneer on calibration based on downsampling. The initial work was based on the HCF-COM feature of Harmsen (2003) (see Section 6.1.1). The problem with HCF-COM is that its variation, even within one class, is enormous, and even though it is statistically greater for natural images than for steganograms, this difference may not be significant. In this situation, calibration may give a baseline for comparison and to improve the discrimination.

Most of the work on downsampling has aimed to identify a single discriminating feature which in itself is able to discriminate between steganograms and clean images. This eliminates the need for a classification algorithm; only a threshold θ needs to be chosen. If f is the discriminating feature, we predict one class label for $f < \theta$ and the alternative class for $f > \theta$. Therefore we will not discuss feature vectors in this section. However, there is no reason why one could not combine a number of the proposed statistics, or even intermediate quantities, into feature vectors for machine learning. We have not seen experiments on such feature vectors, and we leave it as an exercise for the reader.

9.3.1 Downsampling as Calibration

Ker (2005b) suggests downsampling by a factor of two in each dimension. Let $I^{\downarrow}$ denote the down sampled version of an image I. Each pixel of $I^{\downarrow}$ is simply the average of four (2×2) pixels of I, as shown in Figure 9.2. Mathematically we write

$$I^{\downarrow}(x, y) = \left\lfloor \frac{1}{4} \sum_{i,j=0,1} I(2x + i, 2y + j) \right\rfloor .$$

Except for the rounding, this is equivalent to the low-pass component of a (2-D) Haar wavelet decomposition.

Downsampling as calibration is based on the assumption that the embedding distortion $E_{i,j}$ is a random variable, identically and independently distributed for each pixel (i, j). Taking the average of four pixels, we reduce the variance of the distortion. To see this, compare the downsampled pixels of a clean image I_c and a

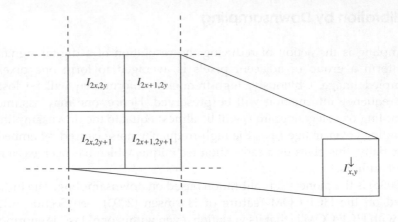

Figure 9.2 Downsampling *á la* Ker

steganogram I_s:

$$I_c^{\downarrow}(x,y) = \left\lfloor \frac{1}{4} \sum_{i,j=0,1} I_c(2x+i, 2y+j) \right\rfloor, \tag{9.2}$$

$$I_s^{\downarrow}(x,y) = \left\lfloor \frac{1}{4} \sum_{i,j=0,1} I_c(2x+i, 2y+j) + \frac{1}{4} \sum_{i,j=0,1} E_{2x+i,2y+j} \right\rfloor. \tag{9.3}$$

If $E_{(x,y)}$ are identically and independently distributed, the variance of $E_{(x,y)}$ is a quarter of the variance of $E_{(x,y)}$. If $E_{(x,y)}$ has zero mean, this translates directly to the distortion power on $I^{\downarrow}$ being a quarter of the distortion power on I.

Intuitively one would thus expect downsampling to work as a cover estimate with respect to most features f. In particular, we would expect that

$$f(I_c) \approx f(I_c^{\downarrow}), \tag{9.4}$$

$$f(I_c) - f(I_s) > f(I_c^{\downarrow}) - f(I_s^{\downarrow}). \tag{9.5}$$

If this holds, a natural image can be recognised as having $f(I) \approx f(I^{\downarrow})$, whereas $f(I_s) < f(I_s^{\downarrow})$ for steganograms.

We shall see later that this intuition is correct under certain conditions, whereas the rounding (floor function) in (9.3) causes problems in other cases. In order to explore this, we need concrete examples of features using calibration. We start with the HCF-COM.

9.3.2 Calibrated HCF-COM

The initial application of downsampling for calibration (Ker, 2005b) aimed to adapt HCF-COM (Definition 4.3.5) to be effective for grey-scale images. We remember that Harmsen's original application of HCF-COM depended on the correlation between colour channels, and first-order HCF-COM features are ineffective on grey-scale images. Downsampling provides an alternative to second-order HCF-COM on grey-scale images. The ratio-calibrated HCF-COM feature used by Ker is defined as

$$f_1(I) = \frac{C(I)}{C(I^\downarrow)},$$

where $C(I)$ is the HCF-COM of image I. Ker's experiment showed that f_1 had slightly better accuracy than C against LSB± at 50% of capacity, and significantly better at 100% of capacity. In Figure 9.3 we show how the HCF-COM features vary with the embedding rate for a number of images. Interestingly, we see that for some, but not all images, f_1 show a distinct fall around 50–60% embedding.

In Figure 9.3(b), we compare f_1, $C(I)$ and $C(I^\downarrow)$ for one of the images where f_1 shows a clear fall. We observe that neither $C(I)$ nor $C(I^\downarrow)$ shows a similar dependency on the embedding rate, confirming the usefulness of calibration. However, we can also see that downsampling does not at all operate as a cover estimate, as $C(I^\downarrow)$ can change more with the embedding rate than $C(I)$ does.

Ker (2005b) also considered the adjacency HCF-COM, as we discussed in Section 4.3. The calibrated adjacency HCF-COM is defined as the scalar feature f_2 given as

$$f_2(I) = \frac{C_2(I)}{C_2(I^\downarrow)}, \quad \text{where}$$

$$C_2(M) = \frac{\sum_{i,j=0}^{n}(i+j)\,|H_M^2(i,j)|}{\sum_{i,j=0}^{n}|H_M^2(i,j)|}, \quad \text{as in (4.7).}$$

Our experiments with uncompressed images in Figure 9.4 show a more consistent trend than we had with the regular HCF-COM.

9.3.3 The Sum and Difference Images

Downsampling as calibration works well in some situations, where it is a reasonable cover estimate. In other situations, it is a poor cover estimate, even with respect to the features f_1 and f_2. We will have a look at when this happens, and how to amend the calibration technique to be more robust.

We define

$$Y_{x,y} = \sum_{i,j=0,1} I_c(2x+i, 2y+j),$$

$$X_{x,y} = \sum_{i,j=0,1} E_{2x+i,2y+j}.$$

Clearly, we can rewrite (9.2) and (9.3) defining the downsampled images as $I_c^\downarrow(x,y) = \lfloor Y_{x,y}/4 \rfloor$ and $I_s^\downarrow(x,y) = \lfloor (Y_{x,y} + X_{x,y})/4 \rfloor$.

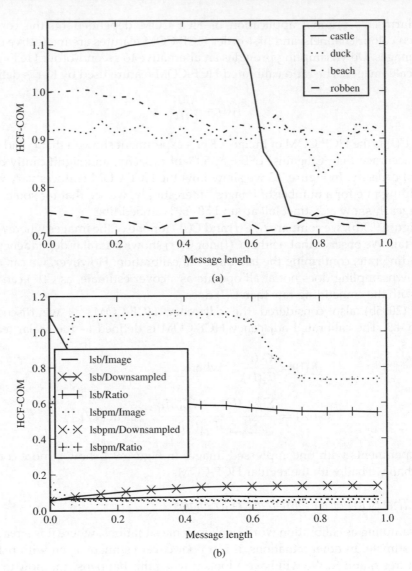

Figure 9.3 HCF-COM: (a) calibrated for various images; (b) calibrated versus non-calibrated

The critical question is the statistical distribution of $Y_{x,y} \bmod 4$. With a uniform distribution, both f_1 and f_2 will be close to 1 for natural images and significantly less for steganograms. Assuming that $Y_{x,y} \bmod 4$ is uniformly distributed, Ker (2005a) was able to prove that

$$f(I_c) - f(I_s) > f(I_c^{\downarrow}) - f(I_s^{\downarrow}),$$

and it was verified empirically that

$$f(I_c) \approx f(I_c^{\downarrow}).$$

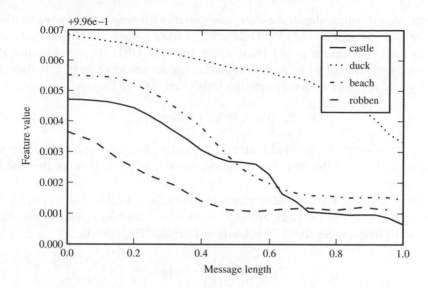

Figure 9.4   Calibrated adjacency HCF-COM for various images

According to Ker, uniform distribution is typical for scanned images. However, images decompressed from JPEG would tend to have disproportionately many groups for which $Y_{x,y} \equiv 0 \pmod 4$.

We can see that if $Y_{x,y} \equiv 0 \pmod 4$, a positive embedding distortion $X_{x,y} = 1, 2, 3$ will disappear in the floor function in the definition of $I_s^{\downarrow}(x, y)$, while negative distortion $X_{x,y} = -1, -2, -3$ will carry through. This means that the embedding distortion $I_s^{\downarrow}(x, y) - I_c^{\downarrow}(x, y)$ is biased, with a negative expectation. It can also cause the distortion to be stronger in $I^{\downarrow}$ than in I, and not weaker as we expected. The exact effect has not been quantified in the literature, but the negative implications can be observed on some of the proposed features using calibration. They do not give good classifiers for previously compressed images.

The obvious solution to this problem is to avoid the rounding function in the definition $I^{\downarrow}$:

$$I^{\downarrow}(x, y) = \left\lfloor \frac{1}{4} \sum_{i,j=0,1} I_{2x+i, 2y+j} \right\rfloor .$$

In fact, the only reason to use this definition is to be able to treat $I^{\downarrow}$ as an image of the same colour depth as I. However, there is no problem using the sum image

$$I^S(x, y) = \sum_{i,j=0,1} I_{2x+i, 2y+j}$$

as the calibrated image. The only difference between $I^{\downarrow}$ and I^s is that the latter has four times the colour depth, that is a range $0 \ldots 1020$ if I is an 8-bit image.

The increased colour depth makes I_S computationally more expensive to use than $I^{\downarrow}$. If we want to calculate HAR3D-3, using a joint histogram across three colour channels and calculating a 3-D HCF using the 3-D DFT, the computational cost increases by a factor of 64 or more. A good compromise may be to calculate a sum image by adding pairs in one dimension only (Ker, 2005a), that is

$$I^s(x,y) = I(2x,y) + I(2x+1,y).$$

Thus the pixel range is doubled instead of quadrupled, saving some of the added computational cost of the DFT. For instance, when a 3-D DFT is used, the cost factor is 8 instead of 64.

The HCF of I^s will have twice as many terms as that of I, because of the increased pixel range. In order to get comparable statistics for I^s and for I, we can use only the lower half of frequencies for I^s. This leads to Ker's (2005a) statistic

$$f_3(I) = \frac{C(I)}{C'(I^s)}, \quad \text{where}$$

$$C'(I^s) = \frac{\sum_{k=0}^{N/2} k\,|H_{I^s}(k)|}{\sum_{k=0}^{N/2} |H_{I^s}(k)|},$$

with $N = 256$ for an 8-bit image I. It is the high-frequency components of the HCF of I^s that are discarded, meaning that we get rid of some high-frequency noise. There is no apparent reason why f_3 would not be applicable to grey-scale images, but we have only seen it used with colour images as discussed below in Section 9.3.4.

The discussion of 2-D histograms and difference matrices in Chapter 4 indicate that the greatest effect of embedding can be seen in the differences between neighbour pixels, rather than individual pixels or even pixel pairs. Pixel differences are captured by the high-pass Haar transform. We have already seen the sum image I^s, which is a a low-pass Haar transform across one dimension only. The difference image can be defined as

$$I^d(x,y) = I(2x,y) - I(2x+1,y) + 255,$$

and it is a high-pass Haar transform across one dimension. Li *et al.* (2008a) suggested using both the difference and sum images, using the HCF-COM $f_1(I^d)$ and $f_1(I^s)$ as features. Experimentally, Li *et al.* show that $f_1(I^d)$ is a better detector than both $f_1(I^s)$ and $f_1(I)$.

The difference matrix $D_{1,0}$ differs from the difference matrix of Chapter 4 in two ways. Firstly, the range is adjusted to be $0 \ldots 510$ instead of ± 255. Secondly, it is downsampled by discarding every other difference; we do not consider $I(2x-1,y) - I(2x,y) + 255$. The idea is also very similar to wavelet analysis, except that we make sure to use an integer transform and use it only along one dimension.

Since the sum and difference images I^s and I^d correspond respectively to the low-pass and high-pass Haar wavelets applied in one dimension only, the discussion

above, combined with the principles of Cartesian calibration, may well be used to
justify features calculated from the wavelet domain.

9.3.4 Features for Colour Images

In the grey-scale case we saw good results with a two-dimensional adjacency HCF-
COM. If we want to use this in the colour case, we can hardly simultaneously
consider the three colour channels jointly, because of the complexity of a 6-D Fourier
transform. One alternative would be to deal with each colour channel separately, to
get three features.

Most of the literature on HCF-COM has aimed to find a single discriminating
feature, and to achieve this Ker (2005a) suggested adding all the three colour
components together and then taking the adjacency HCF-COM. Given an RGB image
I, this gives us the *totalled image*

$$I^t(x,y) = I^R(x,y) + I^G(x,y) + I^B(x,y).$$

This can be treated like a grey-scale image with three times the usual pixel range,
and the usual features $f_i(I^t)$ for $i = 1, 2, 3$ as features of I.

The final detector recommended by Ker is calculated using the totalled image and
calibrated with the sum image:

$$f_4(I) = \frac{C_2'(I^t)}{C_2'(I^{ts})}, \quad \text{where}$$

$$C_2'(J) = \frac{\sum_{i=0}^{n} \sum_{j=0}^{n} (i+j) |H_J(i,j)|}{\sum_{i=0}^{n} \sum_{j=0}^{n} |H_J(i,j)|},$$

where $n = 3N/2$ is half the pixel range of I^t and N the pixel range of I.

9.3.5 Pixel Selection

The premise of calibrated HCF-COM is that $C(I_c) \approx C(I_c^{\downarrow})$. The better this approximation is, the better we can hope the feature to be. Downsampling around an edge, i.e.
where the 2×2 pixels being merged have widely different values, may create new
colours which were not present in the original image I_c. Li *et al.* (2008b) improved the
feature by selecting only smooth pixel groups from the image.

Define a pixel group as

$$S_{i,j} = \{(2i + i', 2j + j') | i', j' = 0, 1\}.$$

Note that the four co-ordinates of $S_{i,j}$ indicate the four pixels contributing to $I^{\downarrow}(i,j)$.
We define the 'smoothness' of $S_{i,j}$ as

$$D_{i,j} = |I(2i, 2j) - I(2i, 2j + 1)| + |I(2i, 2j + 1) - I(2i + 1, 2j + 1)|$$
$$+ |I(2i + 1, 2j + 1) - I(2i + 1, 2j)| + |I(2i + 1, 2j) - I(2i, 2j)|,$$

and we say that a pixel group $S_{i,j}$ is 'smooth' if $D_{i,j} < T$ for some suitable threshold. Li *et al.* recommend $T = 10$ based on experiments, but they have not published the details.

The *pixel selection* image is defined (Li *et al.*, 2008b) as

$$\text{ps}(I) = \left\{ I(x,y) \,\middle|\, (x,y) \in \bigcup_{D_{i,j}<T} S_{i,j} \right\}.$$

Forming the downsampled pixel-selection of I, we get

$$\text{ps}^{\downarrow}(I) = \left\{ \frac{1}{4} \sum_{(x,y)\in S_{i,j}} I(x,y) \,\middle|\, \forall(i,j), D_{i,j} < T \right\}.$$

Based on these definitions, we can define the calibrated pixel-selection HCF-COM as

$$f_3^{\text{ps}}(I) = \frac{C(\text{ps}(I))}{C(\text{ps}^{\downarrow}(I))}.$$

It is possible to create second-order statistics using pixel selection, but it requires a twist. The 2-D HCF-COM according to Ker (2005b) considers every adjacent pair, including pairs spanning two pixel groups. After pixel selection there would be adjacency pairs formed with pixels that were nowhere near each other before pixel selection. In order to make it work, we need to adapt both the pixel selection and the second-order histogram.

The second-order histogram is modified to count only pixel pairs within a pixel group $S_{i,j}$, i.e.

$$h_2(m,n) = \#\left\{ (2i,j) \,\middle|\, I(2i,j) = m \wedge I(2i+1,j) = n \right\}.$$

The pixel selection formula is modified with an extra smoothness criterion on an adjacent pixel group. To avoid ambiguity, we also define the pixel selection as a set ps_2 of adjacency pairs, so that the adjacency histogram can be calculated directly as a standard (1-D) histogram of ps_2.

We define $\text{ps}_2^{\downarrow}(I)$ first, so that both elements of each pair satisfy the criterion $D_{x,y} < T$. Thus we write

$$\text{ps}_2^{\downarrow}(I) = \left\{ (I_{i,j}^{\downarrow}, I_{i+1,j}^{\downarrow}) | \forall(i,j), D_{i,j} < T \wedge D_{i+1,j} < T \right\}, \text{where}$$

$$I_{i,j}^{\downarrow} = \frac{1}{4} \sum_{(x,y)\in S_{i,j}} I(x,y).$$

We now want to define $ps_2(I)$ to include all adjacent pixel pairs of elements taken from an element $S_{i,j}$ used in $ps_2^{\downarrow}(I)$. Thus we write

$$ps_2(I) = \{(I(2x,y), I(2x+1,y)) \mid \{I(2x,y), I(2x+1,y)\} \subset P\}, \text{where}$$

$$P = \bigcup \left\{ S_{2i,j} \cup S_{2i+1,j} \,\middle|\, D_{i,j} < T \wedge D_{i+1,j} < T \right\}.$$

Let $h_2^{ps}(m,n)$ be the histogram of $ps_2(I)$ and $H_2^{ps}(u,v)$ be its DFT. Likewise, let $h_2^{ps^{\downarrow}}(m,v)$ be the histogram of $ps_2^{\downarrow}(I)$ and $H_2^{ps^{\downarrow}}(m,v)$ its DFT. This allows us to define the pixel selection HCF-COM as follows:

$$C_2^{ps}(I) = \frac{\sum_{u,v=0}^{N/2}(u+v)H_2^{ps}(u,v)}{\sum_{u,v=0}^{N/2} H_2^{ps}(u,v)},$$

$$C_2^{ps^{\downarrow}}(I) = \frac{\sum_{u,v=0}^{N/2}(u+v)H_2^{ps^{\downarrow}}(u,v)}{\sum_{u,v=0}^{N/2} H_2^{ps^{\downarrow}}(u,v)},$$

and the calibrated pixel selection HCF-COM is

$$f_2^{ps}(I) = \frac{C_2^{ps}(I)}{C_2^{ps^{\downarrow}}(I)}.$$

9.3.6 Other Features Based on Downsampling

Li *et al.* (2008b) introduced a variation of the calibrated HCF-COM. They applied ratio calibration directly to each element of the HCF, defining

$$d_I^1(k) = \frac{|H_I(k)|}{|H_{I^{\downarrow}}(k)|}.$$

We can think of this as a sort of calibrated HCF. Assuming that downsampling is a cover estimate with respect to the HCF, we should have $d_I^1(k) \approx 1$ for a cover image I. According to previous arguments, the HCF should increase for steganograms so that $d_I^1(k) > 1$. For this reason, Li *et al.* capped d_I^1 from below, defining

$$d_I^{1*}(k) = \min(d_I^1(k), 1),$$

and the new calibrated feature is defined as

$$D_1(I) = \frac{\sum_{k=0}^{N/2} s_k d_I^{1*}(k)}{\sum_{k=0}^{N/2} s_k},$$

where $s_k \geq 0$ are some weighting parameters. Li *et al.* suggest both $s_k = 1$ and $s_k = k$, and decide that $s_k = k$ gives the better performance.

The very same approach also applies to the adjacency histogram, and we can define an analogous feature as

$$D_2(I) = \frac{\sum_{k,l=0}^{N/2} s_{k,l} d_I^{2*}(k,l)}{\sum_{k=0}^{N/2} s_{k,l}}, \quad \text{where}$$

$$d_I^2(k,l) = \frac{|H_I^2(k,l)|}{\left|H_{I\downarrow}^2(k,l)\right|} \quad \text{and}$$

$$d_I^{2*}(k,l) = \min(d_I^2(k,l), 1),$$

where $s_{k,l} \geq 0$ and Li *et al.* use $s_{k,l} = k + l$.

Both D_1 and D_2 can be combined with other techniques to improve performance. Li *et al.* (2008b) tested pixel selection and noted that it improves performance, as it does for other calibrated HCF-based features. Li *et al.* (2008a) note that the high-frequency elements of the HCF are more subject to noise, and they show that using only the first $N/4 = 64$ elements in the sums for D_1 and D_2 improves detection. Similar improvements can also be achieved for the original calibrated HCF-COM feature $C(I)/C(I^\downarrow)$.

9.3.7 Evaluation and Notes

All the calibrated HCF-COM features in this section have been proposed and evaluated in the literature as individual discriminants, and not as feature sets for learning classifiers. However, there is no reason not to combine them with other feature sets for use with SVM or other classifiers.

There are many variations of these features as well. There is Ker's original HCF-COM and the D_1 and D_2 features of Li *et al.*, each of which is in a 1-D and a 2-D variant. Each feature can be calculated with or without pixel selection, and as an alternative to using all non-redundant frequencies, one can reduce this to 64 or 128 low-frequency terms. One could also try Cartesian calibration $(C(I), C(I^\downarrow))$ (see Section 9.4) instead of the ratio $C(I)/C(I^\downarrow)$.

Experimental comparisons of a good range of variations can be found in Li *et al.* (2008b) and Li *et al.* (2008a), but for some of the obvious variants, no experiments have yet been reported in the literature. The experiments of Li *et al.* (2008b) also give inconsistent results, and the relative performance depends on the image set used. Therefore, it is natural to conclude that when fine-tuning a steganalyser, all of the variants should be systematically re-evaluated.

9.4 Calibration in General

So far we have discussed cover estimates only. We will now turn to other classes of reference transforms. The definitions of ratio- and difference-calibrated feature $F_{\text{cal}}^{\text{D}}(I)$ and $F_{\text{cal}}^{\text{R}}(I)$ from Section 9.1 are still valid.

An obvious alternative to cover estimates would be a *stego estimate*, where the reference transform aims to approximate a steganogram, so that $F(\text{cal}\, I_s) \approx F(I_s) \approx F(\text{cal}\, I_c)$. When the embedding replaces cover data, as it does in LSB replacement and JSteg, we can estimate a steganogram by embedding a new random message at 100% of capacity. The resulting transform will be exactly the same steganogram, regardless of whether we started with a clean image or one containing a hidden message. The old hidden message would simply be erased by the new one. Fridrich *et al.* (2003a) used this transform in RS steganalysis.

Stego estimation is much harder when the distortion of repeated embeddings adds together, as it would in LSB matching or F5. Double embedding with these stego-systems can cause distortion of ± 2 in a single coefficient. Thus the transform image would be a much more distorted version than any normal steganogram.

Both cover and stego estimates are intuitive to interpret and use for steganalysis. With cover estimates, $\mathbf{F}^D_{\text{cal}}(J) \approx 0$ for covers so any non-zero value indicates a steganogram. With stego estimates it is the other way around, and $\mathbf{F}^D_{\text{cal}}(J) \approx 0$ for steganograms.

Kodovský and Fridrich (2009) also discuss other types of reference transforms. The simplest example is a parallel reference, where

$$F(\text{cal}\, x) = F(x) - F^*,$$

for some constant F^*. This is clearly degenerate, as it is easy to see that

$$F^D_{\text{cal}}(x) = F(x) - (F(x) + F^*) = F^*,$$

so that the difference-calibrated feature vector contains no information about whether the image is a steganogram or not.

The *eraser transform* is in a sense similar to stego and cover transforms. The eraser aims to estimate some abstract point which represents the cover image, and which is constant regardless of whether the cover is clean or a message has been embedded. This leads to two requirements (Kodovský and Fridrich, 2009):

$$\mathbf{F}_{\text{cal}}(I_c) \approx \mathbf{F}_{\text{cal}}(I_s) =: \mathbf{F}_w \text{ and}$$

$$\mathbf{F}_w \text{ sufficiently close to } \mathbf{F}(I_c) \text{ and } \mathbf{F}(I_s).$$

The second requirement is a bit loosely defined. The essence of it is that $\mathbf{F}_w$ must depend on the image I_c/I_s. Requiring that the feature vector of the transform is 'close' to the feature vectors of I_c and I_s is just one way of achieving this.

Both stego and cover estimates result in very simple and intuitive classifiers. One class will have $\mathbf{F}^D_{\text{cal}} \approx \mathbf{0}$ (or $\mathbf{F}^R_{\text{cal}} \approx \mathbf{1}$), and the other class will have something different. This simple classifier holds regardless of the relationship between $\mathbf{F}(I_c)$ and $\mathbf{F}(I_s)$, as long as they are 'sufficiently' different. This is not the case with the eraser transform. For the eraser to provide a simple classifier, the shift $\mathbf{F}(I_s) - \mathbf{F}(I_c)$ caused by embedding must be consistent.

The last class of reference transforms identified by Kodovský and Fridrich (2009) is the *divergence transform*, which serves to boost the difference between clean image and steganogram by pulling $\mathbf{F}(I_c)$ and $\mathbf{F}(I_s)$ in different directions. In other words, $\mathbf{F}_{cal}(I_c) = \mathbf{F}(I_c) - \mathbf{F}_1$ and $\mathbf{F}_{cal}(I_s) = \mathbf{F}(I_s) - \mathbf{F}_2$, where $\mathbf{F}_1 \neq \mathbf{F}_2$.

It is important to note that the concepts of cover and stego estimates are defined in terms of the feature space, whereas the reference transform operates in image space. We never assume that cal $I \approx I_s$ or cal $I \approx I_c$; we are only interested in the relationship between $\mathbf{F}(I)$ and $\mathbf{F}(I_s)$ or $\mathbf{F}(I_c)$. Therefore the same reference transform may be a good cover estimate with respect to one feature set, but degenerate to parallel reference with respect to another.

9.5 Progressive Randomisation

Over-embedding with a new message has been used by several authors in non-learning, statistical steganalysis. Rocha (2006) also used it in the context of machine learning; see Rocha and Goldenstein (2010) for the most recent presentation of the work. They used six different calibrated images, over-embedding with six different embedding rates, namely

$$p \in P = \{1\%, 5\%, 10\%, 25\%, 50\%, 75\%\}.$$

Each p leads to a reference transform $r_p : I \mapsto I_p$ by over-embedding with LSB at a rate p of capacity, forming a steganogram I_p.

Rocha used the reciprocal of the ratio-calibrated feature discussed earlier. That is, for each base feature f, the calibrated feature is given as

$$f_p(I) = \frac{f(I_p)}{f(I)},$$

where I is the intercepted image.

This calibration method mainly makes sense when we assume that intercepted steganograms will have been created with LSB embedding. In a sense, the reference transform r_p will be a stego estimate with respect to any feature, but the hypothetical embedding rate in the stego estimate will not be constant. If we have an image of size N with rN bits embedded with LSB embedding, and we then over-embed with pN bits, the result is a steganogram with qN embedded bits, where

$$E(q) = r + (1 - r)p = r + p - rp. \tag{9.6}$$

This can be seen because, on average, rpN of the new bits will just overwrite parts of the original message, while $(1 - r)pN$ bits will use previously unused capacity. Thus the calibration will estimate a steganogram at some embedding rate q, but q depends not only on p, but also on the message length already embedded in the intercepted image.

Looking at (9.6), we can see that over-embedding makes more of a difference to a natural image I_c than to a steganogram I_s. The difference is particularly significant if I_s has been embedded at a high rate and the over-embedding will largely touch pixels already used for embedding.

Even though Rocha did not use the terminology of calibration, their concept of progressive randomisation fits well in the framework. We note that the concept of a stego estimate is not as well-defined as we assumed in the previous section, because steganograms with different message lengths are very different. Apart from this, progressive randomisation is an example of a stego estimate. No analysis was made to decide which embedding rate p gives the best reference transform r_p, leaving this problem instead for the learning classifier.

We will only give a brief summary of the other elements making up Rocha's feature vector. They used a method of dividing the image into sub-regions (possibly overlapping), calculating features from each region. Two different underlying features were used. The first one is the χ^2 statistic which we discussed in Chapter 2. The second one is a new feature, using Maurer's (1992) method to measure the randomness of a bit string. Maurer's measure is well suited for bit-plane analysis following the ideas from Chapter 5. If we consider the LSB plane as a 1-D vector, it should be a random-looking bit sequence for a steganogram, and may or may not be random-looking for a clean image, and this is what Maurer's test is designed to detect.

Part III
Classifiers

10
Simulation and Evaluation

In Chapter 3 we gave a simple framework for testing and evaluating a classifier using accuracy, and with the subsequent chapters we have a sizeable toolbox of steganalysers. There is more to testing and evaluation than just accuracy and standard errors, and this chapter will dig into this problem in more depth. We will look at the simulation and evaluation methodology, and also at different performance heuristics which can give a more detailed picture than the accuracy does.

10.1 Estimation and Simulation

One of the great challenges of science and technology is *incomplete information*. Seeking knowledge and understanding, we continuously have to make inferences from a limited body of data. Evaluating a steganalyser, for instance, we desire information about the performance of the steganalyser on the complete *population* of images that we might need to test in the future. The best we can do is to select a *sample* (test set) of already known images to check. This sample is limited both by the availability of images and by the time available for testing.

Statistics is the discipline concerned with quantitative inferences drawn from incomplete data. We will discuss some of the basics of statistics while highlighting concepts of particular relevance. Special attention will be paid to the limitations of statistical inference, to enable us to take a critical view on the quantitative assessments which may be made of steganalytic systems. For a more comprehensive introduction to statistics, we recommend Bhattacharyya and Johnson (1977).

10.1.1 The Binomial Distribution

Let us start by formalising the concept of accuracy and the framework for estimating it. In testing a classifier, we are performing a series of independent experiments or trials. Each trial considers a single image, and it produces one out of two possible

Machine Learning in Image Steganalysis, First Edition. Hans Georg Schaathun.
© 2012 John Wiley & Sons, Ltd. Published 2012 by John Wiley & Sons, Ltd.

outcomes, namely success (meaning that the object was correctly classified) or failure (meaning that the object was misclassified).

Such experiments are known as Bernoulli trials, defined by three properties:

1. Each trial has two possible outcomes, often labelled 'success' (S) and 'failure' (F).
2. Every trial in the series has the same probabilities $p = P(S)$ and $P(F) = 1 - p$.
3. The trials are independent, so that the probabilities $P(S)$ and $P(F)$ are invariant, and do not change as a function of previous outcomes.

Evaluating a classifier according to the framework from Chapter 3, we are interested in the probability of success, or correct classification. Thus the accuracy is equal to the success probability, i.e. $p = A$.

A Bernoulli trial can be any kind of independent trial with a binary outcome. The coin toss, which has 'head' (S) or 'tail' (F) as the two possible outcomes, is the fundamental textbook example. Clearly the probability is $p = 50\%$, unless the coin is bent or otherwise damaged. In fact, we tend to refer to any Bernoulli trial with $p = 50\%$ as a coin toss. Other examples include:

- testing a certain medical drug on an individual and recording if the response is positive (S) or negative (F);
- polling an individual citizen and recording if he is in favour of (S) or against (F) a closer fiscal union in the EU;
- polling an individual resident and recording whether he/she is unemployed (S) or not (F);
- catching a salmon and recording whether it is infected with a particular parasite (S) or not (F).

Running a series of N Bernoulli trials and recording the number of successes X gives rise to the binomial distribution. The stochastic variable X can take any value in the range $\{0, 1, \ldots, N\}$, and the probability that we get exactly x successes is given as

$$P(X = x) = \frac{N!}{x!(N-x)!} p^x (1-p)^{N-x} = \binom{N}{x} p^x (1-p)^{N-x};$$

and the expected number of successes is

$$E(X) = N \cdot p.$$

The test phase in the development of a steganalyser is such a series of Bernoulli trials. With a test set of N images, we expect to get $N \cdot A$ correct classifications on average, but the outcome is random so we can get more and we can get less. Some types of images may be easier to classify than others, but since we do not know which ones are easy, this is just an indistinguishable part of the random process.

10.1.2 Probabilities and Sampling

Much of our interest centres around probabilities; the most important one so far being the accuracy, or probability of correct classification. We have used the word probability several times, relying on the user's intuitive grasp of the concept. An event which has a high probability is expected to happen often, while one with low probability is expected to be rare. In reality, however, probability is an unusually tricky term to define, and definitions will vary depending on whether you are a mathematician, statistician, philosopher, engineer or layman.

A common confusion is to mistake probabilities and rates. Consider some Bernoulli trial which is repeated N times and let X denote the number of successes. The success *rate* is defined as $R = X/N$. In other words, the rate is the outcome of an experiment. Since X is a random variable, we will most likely get a different value every time we run the experiment.

It is possible to interpret probability in terms of rates. The success probability p can be said to be the hypothetical success rate one would get by running an infinite number of trials. In mathematical terms, we can say that

$$p = \lim_{N \to \infty} \frac{X}{N}.$$

Infinity is obviously a point we will never reach, and thus no experiment will ever let us measure the exact probability. Statistics is necessary when we seek information about a population of objects which is so large that we cannot possibly inspect every object, be it infinite or large but finite. If we could inspect every object, we would merely count them, and we could measure precisely whatever we needed to know.

In steganalysis, the population of interest is the set $\mathcal{P}$ of all images we might intercept. All the images that Alice sends and Wendy inspects are drawn randomly from $\mathcal{P}$ according to some probability distribution $P_{\mathcal{P}}$, which depends on Alice's tastes and interests. The accuracy A is the probability that the classifier will give the correct prediction when it is presented with an image drawn from $P_{\mathcal{P}}$.

Trying to estimate A, or any other parameter of $\mathcal{P}$ or $P_{\mathcal{P}}$, the best we can do is to study a sample $\mathcal{S} \subset \mathcal{P}$. Every parameter of the population has an analogue parameter defined on the sample space. We have the population accuracy or true accuracy, and we have a sample accuracy that we used as an estimate for the population accuracy. We can define these parameters in terms of the sample and the population. Let $\mathcal{A}$ be a steganalyser. For any image I, we write $\mathcal{A}(I) \in \{C, S\}$ for the predicted class of I and $\mathcal{C}(I) \in \{C, S\}$ for the true class. We can define the population accuracy as

$$A = \sum_{I \in \mathcal{P}} P_{\mathcal{P}}(I)\delta(\mathcal{A}(I) = \mathcal{C}(I)),$$

where $\delta(E) = 1$ if the statement E is true and 0 otherwise. The sample accuracy is defined as

$$\hat{A} = \frac{\#\{I \mid I \in \mathcal{S}, \mathcal{A}(I) = \mathcal{C}(I)\}}{\#\mathcal{S}}.$$

If the sample is representative for the probability distribution on the population, then $\hat{A} \approx A$. But here is one of the major open problems in steganalysis. We know almost nothing about what makes a representative sample of images. Making a representative sample in steganalysis is particularly hard because the probability distribution $P_{\mathcal{P}}$ is under adversarial control. Alice can choose images which she knows to be hard to classify.

10.1.3 Monte Carlo Simulations

It would be useful if we could test the steganalyser in the field, using real steganograms exchanged between real adversaries. Unfortunately, it is difficult to get enough adversaries to volunteer for the experiment. The solution is to use simulations, where we aim to simulate, or mimic, Alice's behaviour.

Simulation is not a well-defined term, and it can refer to a range of activities and techniques with rather different goals. Our interest is in simulating random processes, so that we can synthesise a statistical sample for analysis. Instead of collecting real statistical data in the field, we generate pseudo-random data which just simulates the real data. This kind of simulation is known as *Monte Carlo simulation*, named in honour of classic applications in simulating games.

Our simulations have a lot in common with applications in communications theory and error control coding, where Monte Carlo simulations are well established and well understood as the principal technique to assess error probabilities (Jeruchim *et al.*, 2000). The objective, both in error control coding and in steganalysis, is to estimate the error probability of a system, and we get the estimate by simulating the operating process and counting error events.

Monte Carlo simulations are designed to generate pseudo-random samples from some probability distribution, and this is commonly used to calculate statistics of the distribution. The probability distribution we are interested in is that of classification errors in the classifier, that is a distribution on the sample space {correct, error}, but without knowledge of the error probability we cannot sample it directly. What we can do is use Monte Carlo processes to generate input (images) for the classifier, run the classifier to predict a class label and then compare it to the true label. A sample Monte Carlo system is shown in Figure 10.1.

The input to the classifier is the product of several random processes, or unknown processes which can only be modelled as random processes. Each of those random processes poses a Monte Carlo problem; we need some way to generate pseudo-random samples. The key will usually be a random bit string, so that poses no challenge. The message would also need to be random or random-looking, assuming that Alice compresses and/or encrypts it, but the message length is not, and we know little about the distribution of message lengths. Then we have Alice's choice of whether to transmit a clean image or a steganogram, which gives rise to the class skew. This is another unknown parameter. The cover image comes from some cover source chosen by Alice, and different sources have different distributions.

The figure illustrates the limitations of straight-forward Monte Carlo simulations. We still have five input processes to simulate, and with the exception of the key,

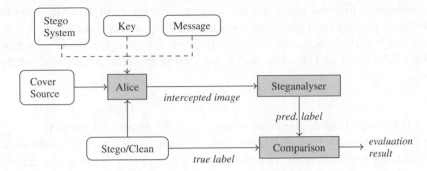

Figure 10.1 Monte Carlo simulator system for steganalysis. All the white boxes are either random processes or processes which we cannot model deterministically

we have hardly any information about their distribution. By pseudo-randomly simulating the five input processes, for given message length, cover source, stego algorithm and class skew, Monte Carlo simulations can give us error probabilities for the assumed parameters. We would require real field data to learn anything about those input distributions, and no one has attempted that research yet.

10.1.4 Confidence Intervals

Confidence intervals provide a format to report estimates of some parameter θ, such as an accuracy or error probability. The confidence interval is a function $(\alpha(D), \beta(D))$ of the observed data D, and it has an associated confidence level p. The popular, but ambiguous, interpretation is that the true value of θ is contained in the confidence interval with probability p. Mathematically, we write it as

$$P(\alpha(D) \leq \theta \leq \beta(D)) = p.$$

The problem with this interpretation is that it does not specify the probability distribution considered. The definition of a confidence interval is that the *a priori* probability of the confidence interval containing θ, when D is drawn randomly from the appropriate probability distribution, is p. Thus both D and θ are considered as random variables.

A popular misconception occurs when the confidence interval (α, β) is calculated for a given observation D, and the confidence level is thought of as a posterior probability. The *a posteriori* probability

$$P(\alpha(D) \leq \theta \leq \beta(D)|D)$$

may or may not be approximately equal to the *a priori* probability $P(\alpha(D) \leq \theta \leq \beta(D))$. Very often the approximation is good enough, but for instance MacKay (2003) gives instructive examples to show that it is not true in general. In order to relate the confidence level to probabilities, it has to be in relation to the probability distribution of the observations D.

The confidence interval is not unique; any interval $(\alpha(D), \beta(D))$ which satisfies the probability requirement is a confidence interval. The most common way to calculate a confidence interval coincides with the definition of error bounds as introduced in Section 3.2. Given an estimator $\bar\theta$ of θ, we define an error bound with confidence level p as

$$C = (\bar\theta - z_{1-p}\sigma, \bar\theta + z_{1-p}\sigma),$$

where σ is the standard deviation of $\bar\theta$ and z_{1-p} is the point such that $P(Z < z_{1-p}) = p$ for a standard normally distributed variable Z. Quite clearly, C satisfies the definition of a confidence interval with a confidence level of p, if $\bar\theta$ is normally distributed, which is the case if the underlying sample of observations is large according to the central limit theorem. In practice, we do not know σ, and replacing σ by some estimate, we get an approximate confidence interval.

Error bounds imply that some estimator has been calculated as well, so that the estimate $\hat\theta$ is quoted with the bounds. A confidence interval, in contrast, is quoted without any associated estimator. Confidence intervals and error bounds can usually be interpreted in the same way. Confidence intervals provide a way to disclose the error bounds, without explicitly disclosing the estimator.

10.2 Scalar Measures

We are familiar with the accuracy A and the error rate $P_e = 1 - A$, as well as the empirical estimators. As a performance indicator, they are limited, and hide a lot of detail about the actual performance of our classifier. Our first step is to see that not all errors are equal.

10.2.1 Two Error Types

Presented with an image I, our steganalyser will return a verdict of either 'stego' or 'non-stego'. Thus we have two possible predicted labels, and also two possible true labels, for a total of four possible outcomes of the test, as shown in the so-called *confusion matrix* (Fawcett, 2006) in Figure 10.2. The columns represent what I actually is, while the rows represent what the steganalyser thinks it is. Both the upper-right and lower-left cells are errors:

False positives (FP) where a natural image (negative) is incorrectly classified as stego (positive).
False negative (FN) where a steganogram is incorrectly classified as a natural image.

In many situations, these two error types may have different associated costs. Criminal justice, for instance, has a very low tolerance for accusing the innocent. In other words, the cost of false positives is considered to be very high compared to that of false negatives.

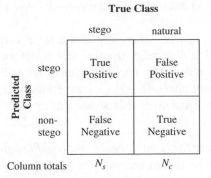

Figure 10.2 The confusion matrix

Let p_{FP} and p_{FN} denote the probability, respectively, of a false positive and a false negative. These are conditional probabilities

$$p_{FP} = P(\text{'stego'}|I \in \mathcal{C}),$$

$$p_{FN} = P(\text{'non-stego'}|I \in \mathcal{S}),$$

where $\mathcal{C}$ and $\mathcal{S}$ are, respectively, the set of all cover images and all steganograms. The detection probability $p_d = 1 - p_{FN}$, as a performance measure, is equivalent to the false negative probability, but is sometimes more convenient. As is customary in statistics, we will use the hat ^ to denote empirical estimates. Thus, $\hat{p}_{FP}$ and $\hat{p}_{FN}$ denote the false positive and false negative rates, respectively, as calculated in the test phase.

The total error probability of the classifier is given as

$$p_e = 1 - A = p_{FP} \cdot P(I \in \mathcal{C}) + p_{FN} \cdot P(I \in \mathcal{S}). \tag{10.1}$$

In other words, the total error probability depends, in general, not only on the steganalyser, but also on the probability of intercepting a steganogram. This is known as the *class skew*. When nothing is known about the class skew, one typically assumes that there is none, i.e. $P(I \in \mathcal{C}) = P(I \in \mathcal{S}) = 0.5$. However, there is no reason to think that this would be the case in steganalysis.

Example 10.2.1 *Provos and Honeyman (2003) ran an experiment to check if image steganography is in active use on eBay and on Usenet news. They used a web crawler to harvest images, and every image found was steganalysed. They found up to 2.1% positives (including false positives) testing for JP Hide on Usenet, and 1% on eBay. These rates are comparable to the false positive rates for the steganalyser, and thus the initial test gives no information as to whether steganography is being used.*

In order to complete the test, they ran a secondary attack, consisting of dictionary attacks aiming to recover the password used by the stego-systems. Such password attacks are time-consuming, and thus constitute a cost of false positives. If the initial steganalyser has a high false positive rate, it will slow down the secondary attack before more candidate images must be tested.

In such a scientific experiment, where the aim was to verify or quantify the use of steganography in the wild, a high false negative rate is not an issue. Unless the number of steganograms in the sample set is negligibly small, we will find a certain number of steganograms, and if we multiply by $(1 - p_{\text{FN}})^{-1}$ we get a good estimate of the total number of steganograms. The FN rate just needs to be so low that we identify a statistically significant number of steganograms.

Example 10.2.2 *Continuing the example above, one could imagine Provos and Honeyman's (2003) system being used by law enforcement or intelligence to identify organised crime applications of steganography. Now, the goal is not to quantify the use, but rather to identify all users, and the FN rate becomes an issue. The FP rate still carries a cost, as false positives will require resources for further investigation, and this cost must be weighted against the risk of missing an important adversary.*

The examples are undoubtedly cases of extreme class skew. Most Internet users have probably not even heard of steganography, let alone considered using it. If, say, $P(I \in S) = 1 - P(I \in C) = 10^{-6}$, it is trivial to make a classifier with error probability $p_e = 10^{-6}$. We just hypothesise 'non-stego' regardless of the input, and we only have an error in the rare case that I is a steganogram. In such a case, the combined error probability p_e is irrelevant, and we would clearly want to estimate p_{FP} and p_{FN} separately.

10.2.2 Common Scalar Measures

Besides p_{FP} and p_{FN}, there is a large number of other measures which are used in the literature. It is useful to be aware of these, even though most of them depend very much on the class skew.

In practice, all measures are based on empirical statistics from the testing; i.e. we find $N_c + N_s$ distinct natural images, and create steganograms out of N_s of them. We run the classifier on each image and count the numbers of false positives (*FP*), true positives (*TP*), false negatives (*FN*) and true negatives (*TN*). Two popular performance indicators are the precision and the recall. We define them as follows:

$$\text{recall} \quad R_1 = \frac{TP}{TP + FN} = 1 - \hat{p}_{\text{FN}},$$

$$\text{precision} \quad P_1 = \frac{TP}{TP + FP}.$$

The precision is the fraction of the predicted steganograms which are truly steganograms, whereas the recall is the number of actual steganograms which are correctly identified. Very often, a harmonic average of the precision and recall is used. This is defined as follows:

$$\text{F-measure} \quad F_1 = 2\frac{P_1 R_1}{P_1 + R_1},$$

which can also be weighted by a constant β. Both $\beta = 2$ and $\beta = 0.5$ are common:

$$F_{\beta}\text{-measure} \qquad F_{\beta} = (1 + \beta^2)\frac{P_1 R_1}{\beta^2 P_1 + R_1}.$$

Finally, we repeat the familiar accuracy:

$$\text{accuracy} \qquad A = \frac{TP + TN}{TP + FN + TN + FP},$$

which is the probability of correct classification. Of all of these measures, only the recall is robust to class skew. All the others use numbers from both columns in the confusion matrix, and therefore they will change when we change the class sizes N_c and N_s relative to each other.

10.3 The Receiver Operating Curve

So far we have discussed the evaluation of discrete classifiers which make a hard decision about the class prediction. Many classifiers return or can return soft information, in terms of a *classification score* $h(\mathbf{x})$ for any object (image) $\mathbf{x}$. The hard decision classifier will then consist of two components; a model, or heuristic function $h(\mathbf{x})$ which returns the classification score, and a threshold w_0. The hard decision rule assigns one class for $h(\mathbf{x}) < w_0$ and the other for $h(\mathbf{x}) > w_0$. Classification scores close to the threshold indicate doubt. If $|h(\mathbf{x}) - w_0|$ is large, the class prediction can be made with confidence.

The threshold w_0 can be varied to control the trade-off between the false negative rate $1 - \hat{p}_{FN}$ and the false positive rate $\hat{p}_{FP}$. When we move the threshold, one will increase and the other will decrease. This means that a soft decision classifier defined by $h(\cdot)$ leaves us some flexibility, and we can adjust the error rates by varying the threshold. *How can we evaluate just the model $h(\cdot)$ universally, without fixing the threshold?*

Figure 10.3 shows an example of the trade-off controlled by the threshold w_0. We have plotted the detection rate $1 - \hat{p}_{FN}$ on the y-axis and the false positive rate $\hat{p}_{FP}$ on the x-axis. Every point on the curve is a feasible point, by choosing a suitable threshold. Thus we can freely choose the false positive rate, and the curve tells us the resulting detection rate, or vice versa. This curve is known as the *receiver operating characteristic* (ROC). We will discuss it further below, starting with how to calculate it empirically.

10.3.1 The libSVM API for Python

Although the command line tools include everything that is needed to train and test a classifier, more advanced analysis is easier by calling libSVM through a general-purpose programming language. The libSVM distribution includes two SVM modules, `svm` and `svmutil`.

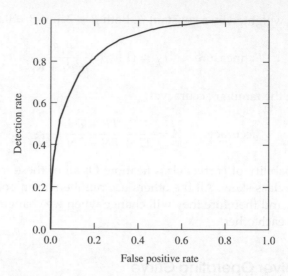

Figure 10.3 Sample ROC curves for the PEV-219 feature vector against 512 bytes F5 embedding. The test set is the same as the one used in Chapter 8

Code Example 10.1 Example of training and testing of a classifier using the libSVM API for Python. The training set is given as a list of labels `labels1` and a list of feature vectors `fvectors1`, where each feature vector is a list of numeric values. The test set is similarly given by `labels2` and `fvectors2`

```
train   = svm.svm_problem( labels1, fvectors1 )
param   = svm.svm_parameter( )
model   = svmutil.svm_train( dataset, param )
(L,A,V) = svmutil.svm_predict( labels2, fvectors2, model )
```

Code Example 10.1 shows how training and testing of an SVM classifier can be done within Python. The output is a tuple (L, A, V). We will ignore A, which includes accuracy and other performance measures. Our main interest is the list of classification scores, V. The list L of predicted labels is calculated from V, with a threshold $w_0 = 0$. Thus a negative classification score represents one class, and a positive score the other.

It may be instructive to see the classification scores visualised as a histogram. Figure 10.4 gives an example of classification scores from two different classifiers using Shi *et al.*'s Markov features. Steganograms (black) and natural images (grey) are plotted separately in the histogram. A threshold is chosen as a point on the x-axis, so that scores to the left will be interpreted as negatives (natural image) and scores to the right as positives (steganograms). Grey bars to the right of the threshold point are false positives, while black bars to the left are false negatives.

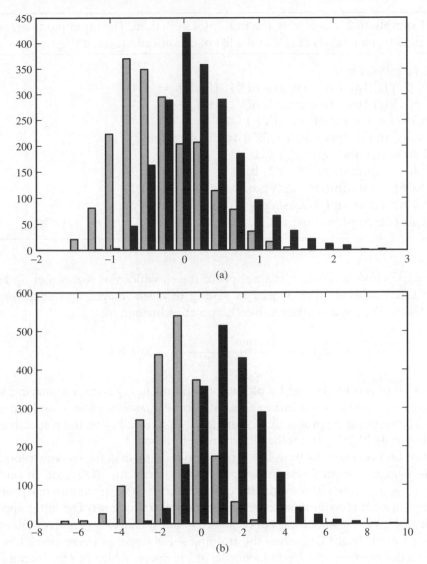

Figure 10.4 Distribution of classification scores for an SVM steganalyser using Shi *et al.*'s Markov model features: (a) default libSVM parameters; (b) optimised SVM parameters

Clearly, we have a trade-off between false positives and false negatives. The further to the left we place our threshold, the more positives will be hypothesised. This results in more detections, and thus a lower false negative rate, but it also gives more false positives. Comparing the two histograms in Figure 10.4(b), it is fairly easy to choose a good threshold, almost separating the steganograms from the natural images. The steganalyser with default parameters has poor separation. Regardless of where we place the threshold, there will be a lot of instances on the wrong side. In this case the trade-off is very important, and we need to prioritise between the two error rates.

Code Example 10.2 Generating the ROC plot in Python. The input parameters are a list of classification labels (±1) Y and a list of classification scores V

```
def rocplot(Y,V):
    V1 = [ V[i] for i in range(len(V)) if Y[i] == −1 ]
    V2 = [ V[i] for i in range(len(V)) if Y[i] == +1 ]
    (mn,mx) = ( min(V), max(V) )
    bins = [ mn + (mx−mn)*i/250 for i in range(251) ]
    (H1,b) = np.histogram( V1, bins )
    (H2,b) = np.histogram( V2, bins )
    H1 = np.cumsum( H1.astype( float ) ) / len(V1)
    H2 = np.cumsum( H2.astype( float ) ) / len(V2)
    matplotlib.pyplot.plot( H1, H2 )
```

The ROC curve is also a visualisation of the classification scores. Let S_C and S_S be the sets of classification scores in testing of clean images and steganograms, respectively. We calculate the relative, cumulative histograms

$$c_X(x) = \frac{\#\{y \mid y \in S_X, y \leq x\}}{\#C_X}, \quad X = S, C.$$

The ROC plot is then defined by plotting the points $(c_S(x), c_C(x))$ for varying values of x. Thus each point on the curve represents some threshold value x, and it displays the two corresponding probability estimates $1 - \hat{p}_{FN}$ and $\hat{p}_{FP}$ on the respective axes. Code Example 10.2 illustrates the algorithm in Python.

It may be necessary to tune the bin boundaries to make a smooth plot. Some authors suggest a more brute-force approach to create the ROC plot. In our code example, we generate 250 points on the curve. Fawcett (2006) generates one plot point for every distinct classification score present in the training set. The latter approach will clearly give the smoothest curve possible, but it causes a problem with vector graphics output (e.g. PDF) when the training set is large, as every point has to be stored in the graphics file. Code Example 10.2 is faster, which hardly matters since the generation of the data set Y and V would dominate, and more importantly, the PDF output is manageable regardless of the length of V.

10.3.2 The ROC Curve

The ROC curve visualises the trade-off between detection rate and false positives. By changing the classification threshold, we can move up and down the curve. Lowering the threshold will lead to more positive classifications, both increasing the detection (true positives) and false positives alike.

ROC stands for 'receiver operating characteristic', emphasising its origin in communications research. The acronym is widely used within different domains, and the origin in communications is not always relevant to the context. It is not unusual to see the C interpreted as 'curve'. The application of ROC curves in machine learning can be traced back at least to Swets (1988) and Spackman (1989), but it has seen an increasing popularity in the last decade. Ker (2004b) interpreted ROC as *region of confidence*, which may be a more meaningful term in our applications. The region under the ROC curve describes the confidence we draw from the classifier.

The ROC space represents every possible combination of FP and TP probabilities. Every discrete classifier corresponds to a point (p_{FP}, p_{TP}) in the ROC space. The perfect, infallible classifier would be at the point $(1, 0)$. A classifier which always returns 'stego' regardless of the input would sit at $(1, 1)$, and one which always returns 'clean' at $(0, 0)$.

The diagonal line, from $(0, 0)$ to $(1, 1)$, corresponds to some naïve classifier which returns 'stego' with a certain probability independently of the input. This is the worst possible steganalyser. If we had a classifier $C(I)$ operating under the main diagonal, i.e. where $1 - p_{FN} < p_{FP}$, we could easily convert it into a better classifier $C'(I)$. We would simply run $C(I)$ and let C' return 'stego' if $C(I)$ returns 'non-stego' and vice versa. The resulting error probabilities p'_{FP} and p'_{FN} of C' would be given as

$$p'_{FP} = 1 - p_{FN},$$
$$1 - p'_{FN} = p_{FP}.$$

This conversion reflects any point on the ROC curve around the main diagonal and allows us to assume that any classifier should be above the main diagonal.

This is illustrated in Figure 10.5, where the worthless classifier is shown as the dashed line. Two classifiers, with and without grid search, are compared, and we clearly see that the one with grid search is superior for all values. We have also shown the reflection of the inferior ROC curve around the diagonal. The size of the grey area can be seen as its improvement over the worthless classifier.

Where discrete classifiers correspond to a point in ROC space, classifiers with soft output, such as SVM, correspond to a curve. Every threshold corresponds to a point in ROC space, so varying the threshold will draw a curve, from $(0, 0)$ corresponding to a threshold of ∞ to $(1, 1)$ which corresponds to a threshold of $-\infty$.

The ROC curve for a soft output classifier is good at visualising the performance independently of the selected threshold. Figure 10.6 shows an example comparing different feature vectors. As we can see, we cannot consistently rank them based on the ROC curve. Two of the curves cross, so that different feature vectors excel at different FP probabilities. Which feature vector we prefer may depend on the use case.

One of the important properties of the ROC curve is that it is invariant irrespective of class skew. This follows because it is the collection of just feasible points (p_{FP}, p_{TP}), and both p_{FP} and p_{TP} measure the performance for a single class of classifier input and consequently are invariant. We also note that classification scores are considered only as ordinal measures, that is, they rank objects in order of how plausibly they could be steganograms. The numerical values of the classification scores are irrelevant to the ROC plots, and thus we can compare ROC plots of different classifiers even if the

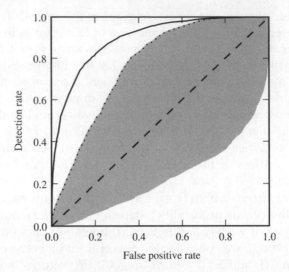

Figure 10.5 Sample ROC curves for the PEV-219 feature vector against 512 bytes F5 embedding. The test set is the same as the one used in Chapter 8. The upper curve is from libSVM with grid search optimised parameters, whereas the lower curve is for default parameters. The shaded area represents the Giri coefficient of the non-optimised steganalyser. The features have been scaled in both cases

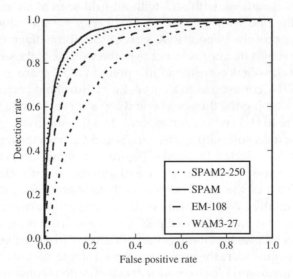

Figure 10.6 ROC plot comparing different feature vectors on a 10% LSB± embedding

classification scores are not comparable. This is important as it allows us to construct ROC plots for arbitrary classifier functions $h(\cdot)$ without converting the scores into some common scale such as likelihoods or p-values. Any score which defines the same order will do.

10.3.3 Choosing a Point on the ROC Curve

There is a natural desire to search for one scalar heuristic to rank all classifiers according to their performance. What the ROC plot tells us is that there is no one such scalar. In Figure 10.6 we clearly see that one classifier may be superior at one point and inferior at another. Every point on the curve matters. Yet sometimes we need a scalar, and different authors have selected different points on the ROC curve for comparison. The following suggestions have been seen in the steganalysis literature, each with good theoretical reasons (Westfeld, 2007):

1. Fix $p_{FP} = p_{FN}$, and use this combined error probability as the performance heuristic.
2. Fix $p_{FP} = 0.01$, and use p_{FN} as the heuristic.
3. Fix $p_{FN} = 0.5$, and use p_{FP} as the heuristic.

Using equal error rates $p_{FP} = p_{FN}$ is very common in for instance biometrics, and has the advantage that the total error probability in (10.1) becomes independent of class skew.

Aiming at a low false positive probability is natural when one targets applications where it is more serious to falsely accuse the innocent than fail to detect the guilty, which is the case in criminal law in most countries. Some early papers in steganalysis used zero false positives as their target point on the ROC curve. This would of course be great, with all innocent users being safe, if we could calculate it, but it is infeasible for two reasons. From a practical viewpoint, we should long have realised that no system is perfect, so even though we can get p_{FP} arbitrarily close to zero, we will never get equality. The theoretical viewpoint is even more absolute. Estimating probabilities close to 0 requires very large samples. As the probability p_{FP} to be estimated tends to zero, the required sample size will tend to infinity. The chosen point on the ROC curve has to be one which can be estimated with a reasonably small error bound. Using $p_{FP} = 0.01$ is quite arbitrary, but it is a reasonable trade-off between low FP probability and feasibility of estimation.

The last proposal, with $p_{FN} = 0.5$, advocated by Ker (2004a), may sound strange. Intuitively, we may find 50% detection unacceptable for most applications. However, if the secret communications extend over time, and the steganalyst acts repeatedly, the steganography will eventually be detected. The advantage of $p_{FN} = 0.5$ is that the threshold T is then equal to the median of the classifier soft output H for steganograms, at which point the detection probability is independent of the variance of H. This removes one unknown from the analysis, and less information is required about the statistical properties of steganograms.

10.3.4 Confidence and Variance

The ROC curves we learnt to draw in Section 10.3.1 are mere snapshots of performance on one single test set, just as a single observation of the sample accuracy $\hat{A}$ would be. There has been a tendency in the machine learning literature to overstate observations from such ROC plots (Fawcett, 2003), by treating better as better without distinguishing between approximately equal and significantly better. In order to make confident conclusions, we need a notion of standard errors or confidence intervals.

Variance and confidence intervals are defined for scalar points, and there is no one canonical way to generalise it for curves. However, since each point visualises two scalars, $\hat{p}_{TP}$ and $\hat{p}_{FP}$, we can calculate confidence intervals or error bounds along each of the axes. This is shown in Figure 10.7, where the vertical and horizontal bars show a 95.4% error bound for $\hat{p}_{TP}$ and $\hat{p}_{FP}$ respectively. The error bounds are calculated using the standard estimator for the standard error of the binomial parameter, namely

$$\hat{p} \pm 2 \cdot \sqrt{\frac{\hat{p}(1 - \hat{p})}{N}},$$

where N is the sample size. Note that there is no correspondence between points on the different curves. Each point corresponds to a given classifier threshold, but since classification scores are not comparable between different soft output classifiers, the thresholds are not comparable either.

Python Tip 10.1 *The plot is generated using the* `matplotlib.pyplot.errorbar` *function.*

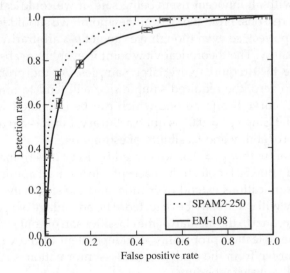

Figure 10.7 Example of ROC plot marking confidence intervals for $\hat{p}_{TP}$ and $\hat{p}_{FP}$ for some thresholds

In the example, we can observe that the confidence intervals are narrow and are far from reaching the other curve except for error rates close to 0. This method of plotting error bars gives a quick and visual impression of the accuracy of the ROC plot, and is well worth including.

Several authors have studied confidence intervals for ROC curves, and many methods exist. Fawcett (2003) gives a broad discussion of the problem, and the importance of considering the variance. Macskassy and Provost (2004) discuss methods to calculate confidence bands for the ROC curve. It is common to use cross-validation or bootstrapping to generate multiple ROC curves, rather than the simple method we described above, but further survey or details on this topic is beyond the scope of this book.

10.3.5 The Area Under the Curve

There is one well-known alternative to picking a single point on the ROC curve, namely measuring the area under the curve (AUC), that is, integrate the ROC curve on the interval $[0, 1]$:

$$\text{AUC} = \int_0^1 p_{\text{TP}} dp_{\text{FP}},$$

where we consider p_{TP} as a function of p_{FP}, as visualised by the ROC curve. The worst possible classifier, namely the random guessing represented by a straight line ROC curve from $(0, 0)$ to $(1, 1)$, has an AUC of 0.5. No realistic classifier should be worse than this. Hence the AUC has range 0.5 to 1.

It may seem more natural to scale the heuristic so that the worst possible classifier gets a value of 0. A popular choice is the so-called *Gini coefficient* (Fawcett, 2003), which is equivalent to the *reliability* ρ that Fridrich (2005) defined in the context of steganalysis. The Gini coefficient can be defined as

$$\rho = 2 \cdot \int_0^1 p_{\text{FP}} dp_{\text{FP}} - 1.$$

The ρ measure corresponds to the shaded area in Figure 10.5. Clearly, the perfect classifier has $\rho = 1$ and random guessing has $\rho = 0$.

The AUC also has another important statistical property. If we draw a random steganogram S and a random clean image C, and let $h(C)$ and $h(S)$ denote their respective classification scores, then

$$\text{AUC} = P(h(C) < h(S)).$$

In other words, the AUC gives us the probability of correct pairwise ranking of objects from different classes.

Westfeld (2007) compared the stability of the four scalar heuristics we have discussed, that is, he sought the measure with the lowest variation when the same

Code Example 10.3 Calculating the AUC. The input parameters are lists of the x and y coordinates of the points on the curve, that is the H1 and H2 variables calculated in Code Example 10.2

```
def auc(H1,H2):
    heights = [ (a+b)/2 for (a,b) in zip(H1[1:],H1[:−1]) ]
    widths  = [ (a−b) for (a,b) in zip(H2[1:],H2[:−1]) ]
    areas   = [ h*w for (h,w) in zip(heights,widths) ]
    return sum(areas)
```

simulation is repeated. He used synthetic data to simulate the soft output from the classifier, assuming either Gaussian or Cauchy distribution. He used 1000 samples each for stego and non-stego, and repeated the experiment to produce 100 000 ROC curves. Westfeld concluded that the error rate $\hat{p} = \hat{p}_{FP} = \hat{p}_{FP}$ at equal error rates and the Gini coefficient ρ were very close in stability, with $p_{FP} = p_{FN}$ being the most stable for Cauchy distribution and ρ being the most stable for Gaussian distribution. The false positive rate $\hat{p}_{FP}$ at 50% detection ($p_{FN} = 0.5$) was only slightly inferior to the first two. However, the detection rate at $p_{FP} = 1\%$ false positives was significantly less stable. This observation is confirmed by our Figure 10.7, where we can see a high variance for the detection rate at low false positive rates, and a low variance for $\hat{p}_{FP}$ when the detection rate is high.

The AUC can be calculated using the trapezoidal rule for numerical integration. For each pair of adjacent points on the ROC curve, (x_i, y_i) and (x_{i+1}, y_{i+1}), we calculate the area of the trapezoid with corners at (x_i, y_i), $(x_i, 0)$, (x_{i+1}, y_{i+1}), $(x_{i+1}, 0)$. The area under the curve is (roughly) made up of all the disjoint trapezoids for $i = 1, 2, \ldots, n - 1$, and we simply add them together. This is shown in Code Example 10.3.

10.4 Experimental Methodology

Experiments with machine learning can easily turn into a massive data management problem. Evaluating a single feature vector, a single stego-system and one cover source is straight forward and gives little reason to think about methodology. A couple of cases can be done one by one manually, but once one starts to vary the cover source, the stego-system, the message length and introduce numerous variations over the feature vectors, manual work is both tedious and error-prone. There are many tests to be performed, reusing the same data in different combinations. A robust and systematic framework to generate, store, and manage the many data sets is required.

Feature extraction is the most time-consuming part of the job, and we clearly do not want to redo it unnecessarily. Thus, a good infrastructure for performing the feature extraction and storing feature data is essential. This is a complex problem, and very few authors have addressed it. Below, we will discuss the most important challenges and some of the solutions which have been reported in the literature.

10.4.1 Feature Storage

The first question is how to store the feature vectors. The most common approach is to use text files, like the sparse format used by the libSVM command line tools. Text files are easy to process and portable between programs. However, text files easily become cumbersome when one takes an interest in individual features and sets out to test variations of feature vectors. Normally, one will only be able to refer to an individual feature by its index, which may change when new features are added to the system, and it is easy to get confused about which index refers to which feature. A data structure that can capture metadata and give easy access to sub-vectors and individual features is therefore useful.

In the experiments conducted in preparation for this book, the features were stored as Python objects, which were stored in binary files using the `pickle` module. This solution is also very simple to program, and the class can be furnished with any kind of useful metadata, including names and descriptions of individual features. It is relatively straight forward to support extraction of sub-vectors. Thus one object can contain all known features, and the different feature vectors of interest can be extracted when needed.

There are problems with this approach. Firstly, data objects which are serialised and stored using standard libraries, like `pickle`, will often give very large files, and even though disk space is cheap, reading and writing to such files may be slow. Secondly, when new features are added to the system, all the files need to be regenerated.

Ker (2004b) proposed a more refined and flexible solution. He built a framework using an SQL database to store all the features. This would tend to take more time and/or skill to program, but gives the same flexibility as the Python objects, and on top of that, it scales better. Ker indicated that more than several hundred million feature vector calculations were expected. Such a full-scale database solution also makes it easy to add additional information about the different objects, and one can quickly identify missing feature vectors and have them calculated.

It may be necessary to store the seeds used for the pseudo-random number generators, to be able to reproduce every object of a simulation exactly. It is clearly preferable to store just the seeds that are used to generate random keys and random messages, rather than storing every steganogram, because the image data would rapidly grow out of hand when the experiments are varied.

10.4.2 Parallel Computation

The feature extraction in the tests we made for Chapters 4–9 took about two minutes per spatial image and more for JPEG. With 5000 images, this means close to a CPU week and upwards. Admittedly, the code could be made more efficient, but programmer time is more expensive than CPU time. The solution is to parallellise the computation and use more CPUs to reduce the running time.

Computer jobs can be parallellised in many different ways, but it always breaks down to one general principle. The work has to be broken into small chunks, so that multiple chunks can be done independently on different CPU cores.

Many chunks will depend on others and will only be able to start when they have received data from other, completed chunks. The differences between paradigms stem from differences in the chunk sizes and the amount of data which must be exchanged when each parallel set of chunks is complete.

For a steganalysis system, a chunk could be the feature extraction from one image. The chunks are clearly independent, so there is no problem in doing them in parallel, and we are unlikely to have more CPU cores than images, so there is no gain in breaking the problem into smaller chunks. This makes parallelism very simple. We can just run each feature extraction as an independent process, dumping the result to file. Thus the computation can easily be distributed over a large network of computers. Results can be accumulated either by using a networked file system, or by connecting to a central database server to record them.

The simplest way to run a large series of jobs on multiple computers is to use a batch queuing system like PBS, Torque or condor. The different jobs, one feature extraction job per image for instance, can be generated and submitted, and the system will distribute the work between computers so that the load is optimally balanced. Batch queue systems are most commonly found on clusters and supercomputers tailored for number crunching, but they can also run on heterogeneous networks. It is quite feasible to install a batch queue system in a student lab or office network to take advantage of idle time out of hours.

There may be good reasons to create a special-purpose queuing system for complex experiments. This can be done using a client–server model, where the server keeps track of which images have been processed and the clients prompt the server for parameters to perform feature extraction. Ker (2004b) recommended a system in which queued jobs were stored in the same MySQL database where the results were stored. This has the obvious advantage that the same system is used to identify missing data and to queue the jobs to generate them.

It is quite fortunate that feature extraction is both the most time-consuming part of the experiment and also simple to parallellise across independent processes. Thus, most of the reward can be reaped with the above simple steps. It is possible to parallellise other parts of the system, but it may require more fine-grained parallelism, using either shared memory (e.g. open-MP) or message passing (e.g. MPI). Individual matrix operations may be speeded up by running constituent scalar operations in parallel, and this would allow us both to speed up feature extractions for one individual image for faster interactive use, and also to speed up SVM or other classifiers. Unless one can find a ready-made parallel-processing API to do the job, this will be costly in man hours, but when even laptop CPUs have four cores or more, this might be the way the world is heading. It can be very useful to speed up interactive trial and error exercises.

Not all programming languages are equally suitable for parallellisation. The C implementation of Python currently employs a global interpreter lock (GIL) which ensures that computation cannot be parallellised. The justification for GIL is that it avoids the overhead of preventing parallel code segments from accessing the same memory. The Java implementation, `Jython`, does not have this limitation, but then it does not have the same libraries, such as `numpy`.

10.4.3 The Dangers of Large-scale Experiments

There are some serious pitfalls when one relies on complex series of experiments. Most importantly, running enough experiments you will always get some interesting, surprising – and false – results. Consider, for instance, the simple case of estimating the accuracy, where we chose to quote twice the standard error to get 95.4% error bounds. Running one test, we feel rather confident that the true accuracy will be within the bound, and in low-risk applications this is quite justified.

However, if we run 20 tests, we would expect to miss the 95.4% error bound almost once on average. Running a thousand tests we should miss the error bound 46 times on average. Most of them will be a little off, but on average one will fall outside three times the standard error. Even more extreme results exist, and if we run thousands of tests looking for a surprising result, we will find it, even if no such result is correct. When many tests are performed in the preparation of analysis, one may require a higher level of confidence in each test. Alternatively, one can confirm results of particular interest by running additional experiments using a larger sample.

The discussion above, of course, implicitly assumes that the tests are independent. In practice, we would typically reuse the same image database for many tests, creating dependencies between the tests. The dependencies will make the analysis less transparent. Quite possibly, there are enough independent random elements to make the analysis correct. Alternatively, it is possible that extreme results will not occur as often, but in larger number when they occur at all.

To conclude, when large suites of experiments are used, care must be taken to ensure the level of confidence. The safest simple rule of thumb is probably to validate surprising results with extra, independent tests with a high significance level.

10.5 Comparison and Hypothesis Testing

In Chapter 3 we gave some rough guidelines on how much different two empirical accuracies had to be in order to be considered different. We suggested that accuracies within two standard errors of each other should be considered (approximately) equal.

A more formal framework to compare two classifiers would be to use a statistical hypothesis test. We remember that we also encountered hypothesis testing for the pairs-of-values test in Chapter 2, and it has been used in other statistical steganalysers as well. It is in fact quite common and natural to phrase the steganalysis problem as a statistical hypothesis test. In short, there are many good reasons to spend some time on hypothesis testing as a methodology.

10.5.1 The Hypothesis Test

A common question to ask in steganalysis research is this: Does steganalyser A have a better accuracy than steganalyser B? This is a question which is well suited for a hypothesis test. We have two hypotheses:

$$H_0: \; A_X = A_Y,$$
$$H_1: \; A_X > A_Y,$$

where A_X and A_X denote the accuracy of steganalyser X and Y respectively. Very often the experimenter hopes to demonstrate that a new algorithm A is better than the current state of the art. Thus H_0 represents the default or negative outcome, while H_1 denotes, in a sense, the desired or positive outcome. The hypothesis H_0 is known as the *null hypothesis*.

Statistical hypothesis tests are designed to give good control of the probability of a false positive. We want to ensure that the evidence is solid before we promote the new algorithm. Hence, we require a known probability model under the condition that H_0 is true, so that we can calculate the probability of getting a result leading us to reject H_0 when we should not. It is less important to have a known probability model under H_1, although we would need it if we also want to calculate the false negative probability.

The test is designed ahead of time, before we look at the data, in such a way that the probability of falsely rejecting H_0 is bounded. The test is staged as a contest between the two hypotheses, with clearly defined rules. Once the contest is started there is no further room for reasoning or discussion. The winning hypothesis is declared according to the rules.

In order to test the hypothesis, we need some observable random variable which will have different distributions under H_0 and H_1. Usually, that is an estimator for the parameters which occur in the hypothesis statements. In our case the hypothesis concerns the accuracies A_X and A_Y, and we will observe the empirical accuracies $\hat{A}_X$ and $\hat{A}_Y$. If $\hat{A}_X \approx \hat{A}_Y$, then H_0 is plausible. If $\hat{A}_X \gg \hat{A}_Y$, it is reasonable to reject H_0. The question is, *how much larger ought $\hat{A}_X$ to be before we can conclude that A is a better steganalyser?*

10.5.2 Comparing Two Binomial Proportions

The accuracy is the success probability of a Bernoulli trial in the binomial distribution, also known as the binomial proportion. There is a standard test to compare the proportions of two different binomial distributions, such as for two different classifiers X and Y. To formulate the test, we rephrase the null hypothesis as

$$H_0 : A_X - A_Y = 0.$$

Thus we can consider the single random variable $Z = \hat{A}_X - \hat{A}_Y$, rather than two variables.

If the training set is at least moderately large, then Z is approximately normally distributed. The expectation and variance of Z are given as follows:

$$E(\hat{A}_X - \hat{A}_Y) = A_X - A_Y, \quad \text{and}$$

$$\text{Var}(\hat{A}_X - \hat{A}_Y) = \frac{A_X(1 - A_X)}{N_X} + \frac{A_Y(1 - A_Y)}{N_Y},$$

which can be estimated as

$$\hat{\sigma}^2 = \frac{\hat{A}_X(1 - \hat{A}_X)}{N_X} + \frac{\hat{A}_Y(1 - \hat{A}_Y)}{N_Y}.$$

Thus we have a known probability distribution under H_0, as required. Under H_0 we expect $Z \approx 0$, while any large value of $|Z|$ indicates that H_0 is less plausible. For instance, we know that the probability $P(|Z| > 2\hat{\sigma}) \approx 4.6\%$. In other words, if we decide to reject H_0 if $|Z| > 2\hat{\sigma}$, we know that the probability of making a false positive is less than 5%. The probability of falsely rejecting the null hypothesis is known as the *significance level* of the hypothesis test, and it is the principal measure of the reliability of the test.

We can transform the random variable Z to a variable with standard normal distribution under H_0, by dividing by the standard deviation. We define

$$Z' = \frac{\hat{A}_X - \hat{A}_Y}{\sqrt{\dfrac{\hat{A}_X(1 - \hat{A}_X)}{N_X} + \dfrac{\hat{A}_Y(1 - \hat{A}_Y)}{N_Y}}}, \tag{10.2}$$

which has zero mean and unit variance under the null hypothesis. The standard normal distribution $N(0, 1)$ makes the observed variable easier to handle, as it can be compared directly to tables for the normal distribution.

So far we have focused on the null hypothesis $A_X = A_Y$, and its negation has implicitly become the alternative hypothesis:

$$H_1' : A_X - A_Y \neq 0.$$

This hypothesis makes no statement as to which of X or Y might be the better, and we will reject H_0 regardless of which algorithm is better, as long as they do not have similar performance.

Very often, we want to demonstrate that a novel classifier X is superior to an established classifier Y. If $A_Y \ll A_X$, it may be evidence that H_0 is implausible, but it hardly proves the point we want to make. In such a case, the alternative hypothesis should naturally be

$$H_1 : A_X - A_Y > 0.$$

Which alternative hypothesis we make has an impact on the decision criterion. We discussed earlier that in a test of H_0 against H_1' at a significance level of 4.6%, we would reject the null hypothesis if $|Z'| > 2$. If we test H_0 against H_1 however, we do not reject the null hypothesis for $Z' < 0$. At a significance level of 4.6% we should reject H_0 if $Z' > 1.69$, since $P(Z > 1.69 \mid H_0) \leq 4.6\%$.

Example 10.5.1 *Suppose we want to check if Cartesian calibration as discussed in Chapter 9 improves the performance of the Pevný features. That is, we want to compare CCPEV-438 (classifier X) with NCPEV-219 (classifier Y). The hypotheses are given as*

$$H_0 : A_X - A_Y = 0,$$

$$H_1 : A_X - A_Y > 0.$$

We observe the random variable Z' as defined in (10.2). Let's choose a significance level of 5% and test the algorithms on a sample of 4000 images. The experiment returns $\hat{A}_X = 0.87125$ and $\hat{A}_Y = 0.8565$, and we calculate

$$Z' = \frac{0.87125 - 0.8565}{\sqrt{\dfrac{0.87125 \cdot (1 - 0.87125)}{4000} + \dfrac{0.8565 \cdot (1 - 0.8565)}{4000}}} \approx 1.9240.$$

We can look up the value $z_{0.95}$ such that $P(Z > z_{0.95}) = 5\%$ for a normally distributed variable $Z \sim N(0, 1)$ in Python:

In [34]: **import** *scipy.stats*

In [35]: N = scipy.stats.norm()

In [36]: N.ppf([0.95])
Out[36]: array([1.64485363])

Here N is defined as a standard normal distribution and ppf() is the inverse cumulative density function. Thus $z_{0.95} = 1.645 < Z'$, and we can reject H_0 at 5% significance. We conclude that Cartesian calibration improves the Pevný features.

10.6 Summary

A much too frequent pitfall in steganalysis is to exaggerate conclusions based on very simplified heuristics. It would be useful to re-evaluate many well-known stego-systems, aiming to get a more balanced, complete and confident assessment. Most importantly, a notion of confidence, or statistical significance, is required. This is well understood in statistics, and in many disciplines it would be unheard of to publish results without an assessment of error bounds and significance. In steganalysis there are still open problems to answer, where the original authors have cut corners.

Furthermore one needs to be aware that different applications will require different trade-offs between false positives and false negatives, and knowledge of the full ROC curve is often necessary. We also know that classifier performance will vary depending on message lengths and cover sources. To get a good picture of field performance, more than just a few combinations must be tried. We will return to these questions in Chapter 14.

11

Support Vector Machines

Support vector machines (SVM) are a good algorithm for starters who enter the field of machine learning from an applied angle. As we have demonstrated in previous chapters, easy-to-use software will usually give good classification performance without any tedious parameter tuning. Thus all the effort can be put into the development of new features.

In this chapter, we will investigate the SVM algorithm in more depth, both to extend it to one-class and multi-class classification, and to cover the various components that can be generalised. In principle, SVM is a linear classifier, so we will start by exploring linear classifiers in general. The trick used to classify non-linear problems is the so-called kernel trick, which essentially maps the feature vectors into a higher-dimensional space where the problem becomes linear. This kernel trick is the second key component, which can generalise to other algorithms.

11.1 Linear Classifiers

An object i is characterised by two quantities, a feature vector $\mathbf{x}_i \in \mathbb{R}^n$ which can be observed, and a class label $y_i \in \{-1, +1\}$ which cannot normally be observed, but which we attempt to deduce from the observed $\mathbf{x}_i$. Thus we have two sets of points in n-space, namely $\{\mathbf{x}_i | y_i = -1\}$ and $\{\mathbf{x}_i | y_i = +1\}$, as illustrated in Figure 11.1. Classification aims to separate these two sets geometrically in $\mathbb{R}^n$.

A linear classifier aims to separate the classes by drawing a hyperplane between them. In $\mathbb{R}^2$ a hyperplane is a line, in $\mathbb{R}^3$ it is a plane and in $\mathbb{R}^n$ it is a subspace of dimension $n - 1$. If the problem is linearly separable, as it is in Figure 11.1(a), it is possible to draw such a hyperplane which cleanly splits the objects into two sets exactly matching the two classes. If the problem is not linearly separable, as in Figure 11.1(b), this is not possible, and any linear classifier will make errors.

Mathematically, a hyperplane is defined as the set $\{\mathbf{x} | \mathbf{x} \cdot \mathbf{w} - w_0 = 0\}$, for some vector $\mathbf{w}$, orthogonal on the hyperplane, and scalar w_0. Using the hyperplane as a

Machine Learning in Image Steganalysis, First Edition. Hans Georg Schaathun.
© 2012 John Wiley & Sons, Ltd. Published 2012 by John Wiley & Sons, Ltd.

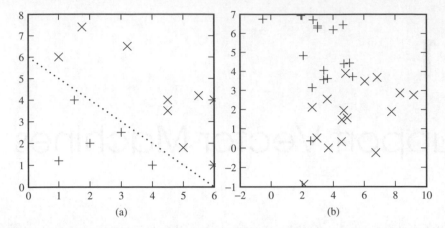

Figure 11.1 Feature vectors visualised in $\mathbb{R}^2$: (a) linearly separable problem; (b) linearly non-separable problem

classifier is straightforward. The function $h(\mathbf{x}) = \mathbf{x} \cdot \mathbf{w} - w_0$ will be used as a classification heuristic. We have $h(\mathbf{x}) < 0$ on one side of the hyperplane and $h(\mathbf{x}) > 0$ on the other, so simply associate the class labels with the sign of $h(\mathbf{x})$. Let y_i and $\mathbf{x}_i$ denote the labels and feature vectors used to calculate the function $h(\cdot)$ during the training phase.

11.1.1 Linearly Separable Problems

The challenge is to find $\mathbf{w}$ to define the hyperplane which gives the best possible classification; but what do we mean by 'best'? Consider the example in Figure 11.2. Which of the two proposed classification lines is the better one?

The solid line has a greater distance to the closest feature vector than the dashed line. We call this distance the margin of the classifier. Intuitively, it sounds like a good idea to maximise this margin. Remembering that the training points are only a random sample, we know that future test objects will be scattered randomly around. They are likely to be relatively close to the training points, but some must be expected to lie closer to the classifier line. The greater the margin, the less likely it is that a random point will fall on the wrong side of the classifier line.

We define two boundary hyperplanes parallel to and on either side of the classification hyperplane. The distance between the boundary hyperplane and the classification hyperplane is equal to the margin. Thus, there is no training point between the two boundary hyperplanes in the linearly separable case.

Consider a point $\mathbf{x}$ and the closest point $\mathbf{z}$ on the classification hyperplane. It is known that the distance $\|\mathbf{x} - \mathbf{z}\|$ between $\mathbf{x}$ and the hyperplane is given as

$$\|\mathbf{x} - \mathbf{z}\| = \frac{|(\mathbf{x} - \mathbf{z}) \cdot \mathbf{w}|}{\|\mathbf{w}\|} = \frac{|h(\mathbf{x}) - h(\mathbf{z})|}{\|\mathbf{w}\|} = \frac{|h(\mathbf{x})|}{\|\mathbf{w}\|},$$

where the first equality is a standard observation of vector algebra.

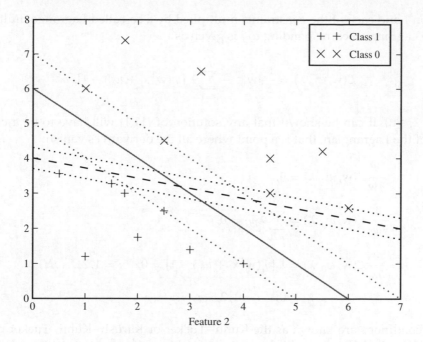

Figure 11.2 Two different classification hyperplanes for the same separable problem

It is customary to scale $\mathbf{w}$ and w_0 so that $h(\mathbf{x}) = \pm 1$ for any point $\mathbf{x}$ on the boundary hyperplanes. The margin is given as $M = \|\mathbf{x} - \mathbf{z}\|$ for an arbitrary point $\mathbf{x}$ on the boundary hyperplanes and its closest point $\mathbf{z}$ on the classification hyperplane. Hence, with scaling, the margin is given as

$$M = \frac{1}{\|\mathbf{w}\|}.$$

To achieve maximum M, we have to minimise $\|\mathbf{w}\|$, that is, we want to solve

$$\min_{\mathbf{w}, w_0} \; J(\mathbf{w}, w_0) = \frac{1}{2} \|\mathbf{w}\|^2, \tag{11.1}$$

$$\text{subject to} \quad y_i(\mathbf{w}^\mathsf{T} \mathbf{x}_i + w_0) \geq 1, \quad i = 1, 2, \dots, N. \tag{11.2}$$

This is a standard quadratic optimisation problem with linear inequality constraints. It is a convex minimisation problem because J is a convex function, meaning that for any $0 \leq t \leq 1$ and any $\mathbf{w}_1, \mathbf{w}_2, w_1, w_2$, we have

$$J(t\mathbf{w}_1 + (1-t)\mathbf{w}_2, tw_1 + (1-t)w_2) \leq tJ(\mathbf{w}_1, w_1) + (1-t)J(\mathbf{w}_2, w_2).$$

Such convex minimisation problems can be solved using standard techniques based on the Lagrangian function $\mathcal{L}(\mathbf{w}, w_0, \boldsymbol{\lambda})$, which is found by adding the cost function J and each of the constraints multiplied by a so-called Lagrange multiplier λ_i. The Lagrangian corresponding to J is given as

$$\mathcal{L}(\mathbf{w}, w_0, \boldsymbol{\lambda}) = \frac{1}{2}\|\mathbf{w}\|^2 - \sum_{i=1}^{N} \lambda_i[y_i(\mathbf{w}^{\mathsf{T}}\mathbf{x}_i + w_0) - 1],$$

where $\lambda_i \geq 0$. It can be shown that any solution of (11.1) will have to be a critical point of the Lagrangian, that is a point where all the derivatives vanish:

$$\frac{\partial}{\partial \mathbf{w}} \mathcal{L}(\mathbf{w}, w_0, \boldsymbol{\lambda}) = \mathbf{0}, \tag{11.3}$$

$$\frac{\partial}{\partial w_0} \mathcal{L}(\mathbf{w}, w_0, \boldsymbol{\lambda}) = \sum_{i=1}^{N} \lambda_i y_i = 0, \tag{11.4}$$

$$\frac{\partial}{\partial \lambda_i} \mathcal{L}(\mathbf{w}, w_0, \boldsymbol{\lambda}) = \lambda_i[y_i(\mathbf{w}^{\mathsf{T}}\mathbf{x}_i + w_0) - 1] = 0, \quad i = 1, 2, \ldots, N, \tag{11.5}$$

$$\lambda_i \geq 0, \quad i = 1, 2, \ldots, N. \tag{11.6}$$

These conditions are known as the Kuhn–Tucker or Karush–Kuhn–Tucker conditions. The minimisation problem can be solved by solving an equivalent problem, the so-called Wolfe dual representation, defined as follows:

$$\max_{\boldsymbol{\lambda}} \quad \mathcal{L}(\mathbf{w}, w_0, \boldsymbol{\lambda}),$$

subject to conditions (11.3) to (11.6), which are equivalent to

$$\mathbf{w} = \sum_{i=1}^{N} \lambda_i y_i \mathbf{x}_i, \tag{11.7}$$

$$\sum_{i=1}^{N} \lambda_i y_i = 0, \tag{11.8}$$

$$\lambda_i \geq 0, \quad i = 1, 2, \ldots, N. \tag{11.9}$$

Substituting for (11.7) and (11.8) in the definition of $\mathcal{L}$ lets us simplify the problem to

$$\max_{\boldsymbol{\lambda}} \sum_{i=1}^{N} \lambda_i - \frac{1}{2} \sum_{i,j} \lambda_i \lambda_j y_i y_j \mathbf{x}_i \cdot \mathbf{x}_j, \tag{11.10}$$

subject to (11.8) and (11.9). Note that the only unknowns are now λ_i. The problem can be solved using any standard software for convex optimisation, such as CVXOPT for Python. Once (11.10) has been solved for the λ_i, we can substitute back into (11.7) to get $\mathbf{w}$. Finally, w_0 can be found from (11.5).

The Lagrange multipliers λ_i are sometimes known as shadow costs, because each represents the cost of obeying one of the constraints. When the constraint is passive, that is satisfied with inequality, $\lambda_i = 0$ because the constraint does not prohibit a solution which would otherwise be more optimal. If, however, the constraint is active, by being met with equality, we get $\lambda_i > 0$ and the term contributes to increasing $\mathcal{L}$, imposing a cost on violating the constraint to seek a better solution. In a sense, we imagine that we are allowed to violate the constraint, but the cost λ does not make it worth while.

The formulation (11.10) demonstrates two key points of SVM. When we maximise (11.10), the second term will vanish for many (i, j) where $\lambda_i = 0$ or $\lambda_j = 0$. A particular training vector $\mathbf{x}_i$ will only appear in the formula for $\mathbf{w}$ if $\lambda_i = 0$, which means that $\mathbf{x}_i$ is on the boundary hyperplane. These training vectors are called *support vectors*.

The second key point observed in (11.10) is that the feature vectors $\mathbf{x}_i$ enter the problem in pairs, and only as pairs, as the inner product $\mathbf{x}_i \cdot \mathbf{x}_j$. In particular, the dimensionality does not contribute explicitly to the cost function. We will use this observation when we introduce the kernel trick later.

11.1.2 Non-separable Problems

In most cases, the two classes are not separable. For instance, in Figure 11.1(b) it is impossible to draw a straight line separating the two classes. No matter how we draw the line, we will have to accept a number of classification errors, even in the training set. In the separable case we used the margin as the criterion for choosing the classification hyperplane. In the non-separable case we will have to consider the number of classification errors and possibly also the size (or seriousness) of each error.

Fisher Linear Discriminant

The Fisher linear discriminant (FLD) (Fisher, 1936) is the classic among classification algorithms. Without diminishing its theoretical and practical usefulness, it is not always counted as a learning algorithm, as it relies purely on conventional theory of statistics.

The feature vectors $\mathbf{x}$ for each class C_i are considered as stochastic variables, with the covariance matrix defined as

$$\Sigma_i^2 = E(\mathbf{x} - \boldsymbol{\mu}_i)(\mathbf{x} - \boldsymbol{\mu}_i)^{\mathrm{T}}, \quad \text{for } \mathbf{x} \in C_i,$$

and the mean as

$$\boldsymbol{\mu}_i = E(\mathbf{x}|\mathbf{x} \in C_i).$$

As for SVM, we aim to find a classifier hyperplane, defined by a linear equation $h(\mathbf{x}) = \mathbf{x} \cdot \mathbf{w} - w_0 = 0$. The classification rule will also be the same, predicting one class for $h(\mathbf{x}) < 0$ and the other for $h(\mathbf{x}) > 0$. The difference is in the optimality criterion. Fisher sought to maximise the difference between $h(\mathbf{x})$ for the different classes, and simultaneously minimise the variation within each class.

The between-class variation is interpreted as the differences between the means of $h(\mathbf{x})$, given as

$$E(h(\mathbf{x})|\mathbf{x} \in C_i) = \sum_{\mathbf{x} \in C_i} \frac{\mathbf{w} \cdot \mathbf{x}}{\#C_i} - w_0 = \mathbf{w} \cdot \boldsymbol{\mu}_i - w_0. \tag{11.11}$$

The variation within a class is given by the variance:

$$\mathrm{Var}(h(\mathbf{x})) = \mathrm{Var}(\mathbf{w}^\mathrm{T}\mathbf{x}) = \mathbf{w}^\mathrm{T} \mathrm{Cov}(\mathbf{x})\mathbf{w} = \mathbf{w}^\mathrm{T}\boldsymbol{\Sigma}_i^2\mathbf{w}. \tag{11.12}$$

Fisher (1936) defined the separation quantitatively as the ratio between the variation between the two classes and the variance within each class. Symbolically, we write the separation as

$$L = \frac{\sigma_{\text{between}}^2}{\sigma_{\text{within}}^2} = \frac{(\mathbf{w}\boldsymbol{\mu}_1 - \mathbf{w}\boldsymbol{\mu}_0)^2}{\mathbf{w}^\mathrm{T}\boldsymbol{\Sigma}_1^2\mathbf{w} + \mathbf{w}^\mathrm{T}\boldsymbol{\Sigma}_{-1}^2\mathbf{w}}.$$

It can be shown that the separation L is maximised by selecting

$$\mathbf{w} = (\boldsymbol{\Sigma}_1^2 + \boldsymbol{\Sigma}_{-1}^2)^{-1}(\boldsymbol{\mu}_1 - \boldsymbol{\mu}_{-1}).$$

The vector $\mathbf{w}$ defines the direction of the hyperplane. The final step is to determine w_0 to complete the definition of $h(\mathbf{x})$. It is natural to choose a classifier hyperplane through a point midway between the two class means $\boldsymbol{\mu}_{-1}$ and $\boldsymbol{\mu}_{+1}$. This leads to

$$w_0 = \frac{\mathbf{w} \cdot (\boldsymbol{\mu}_{-1} + \boldsymbol{\mu}_{+1})}{2},$$

which completes the definition of FLD. More details can be found in standard textbooks on multivariate statistics, such as Johnson and Wichern (1988).

An important benefit of FLD is the computational cost. As we have seen above, it only requires the calculation of the two covariance matrices, and a few linear operations thereon. In practical implementations, we substitute sample mean and sample covariance for $\boldsymbol{\mu}_i$ and $\boldsymbol{\Sigma}_i^2$.

The SVM Discriminant

In SVM there are two quantities which determine the quality of the classification hyperplane. As in the linearly separable case, we still maintain two boundary hyperplanes at a distance M on either side of the classification hyperplane and parallel to it, and we want to maximise M. Obviously, some points will then be misclassified (cf. Figure 11.3), and for every training vector $\mathbf{x}_i$ which falls on the wrong side of the relevant boundary plane, we define the error ϵ_i as the distance from $\mathbf{x}_i$ to the boundary plane. Note that according to this definition, an

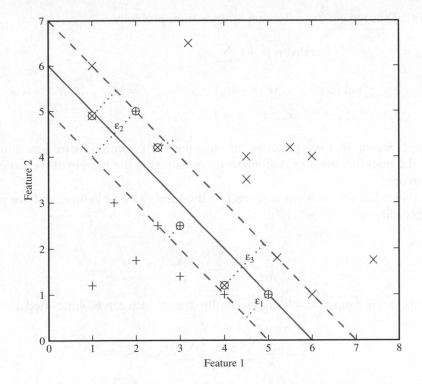

Figure 11.3   Classification with errors

error is not necessarily misclassified. Points which fall between the two boundary hyperplanes are considered errors, even if they still fall on the right side of the classifier hyperplane.

We build on the minimisation problem from the linearly separable case, where we only had to minimise $\|\mathbf{w}\|^2$. Now we also need to minimise the errors ϵ_i, which introduce slack in the constraints, that now become

$$y_i(\mathbf{w} \cdot \mathbf{x} + w_0) \geq 1 - \epsilon_i, \quad i = 1, 2, \ldots, N.$$

The errors ϵ_i are also known as *slack variables*, because of their role in the constraints. For the sake of generality, we define $\epsilon_i = 0$ for every training vector $\mathbf{x}_i$ which falls on the right side of the boundary hyperplanes. Clearly $\epsilon_i \geq 0$ for all i. Training vectors which are correctly classified but fall between the boundaries will have $0 \leq \epsilon_i \leq 1$, while misclassified vectors have $\epsilon \geq 1$.

Minimising $\|\mathbf{w}\|^2$ and minimising the errors are conflicting objectives, and we need to find the right trade-off. In SVM, we simply add the errors to the cost function from the linearly separable case, multiplied by a constant scaling factor C to control the trade-off. Thus we get the following optimisation problem:

$$\min \|\mathbf{w}\|^2 + C \sum_i \epsilon_i,$$

$$\text{subject to} \quad y_i(\mathbf{w} \cdot \mathbf{x} + w_0) \geq 1 - \epsilon_i, \quad i = 1, 2, \ldots, N,$$

$$\epsilon_i \geq 0, \quad i = 1, 2, \ldots, N.$$

A large C means that we prioritise minimising the errors and care less about the size of the margin. Smaller C will focus on maximising the margin at the expense of more errors.

The minimisation problem is solved in the same way as before, and we get the same definition of $\mathbf{w}$ from (11.7):

$$\mathbf{w} = \sum_{i=1}^{N} \lambda_i y_i \mathbf{x}_i,$$

where the λ_i are found by solving the Wolfe dual, which can be simplified as

$$\max_{\lambda} \sum_{i=1}^{N} \lambda_i - \frac{1}{2} \sum_{i,j} \lambda_i \lambda_j y_i y_j \mathbf{x}_i \cdot \mathbf{x}_j, \qquad (11.13)$$

$$\text{subject to} \quad \sum_{i=1}^{N} \lambda_i y_i = 0, \qquad (11.14)$$

$$0 \leq \lambda_i \leq C, \quad i = 1, 2, \ldots, N. \qquad (11.15)$$

It can be shown that, for all i, $\epsilon_i > 0$ implies $\lambda_i = C$. Hence, training vectors which fall on the wrong side of the boundary hyperplane will make the heaviest contributions to the cost function.

11.2 The Kernel Function

The key to the success of SVM is the so-called kernel trick, which allows us to use the simple linear classifier discussed so far, and still solve non-linear classification problems. Both the linear SVM algorithm that we have discussed and the kernel trick are 1960s work, by Vapnik and Lerner (1963) and Aizerman *et al.* (1964), respectively. However, it was not until 1992 that Boser *et al.* (1992) combined the two. What is now considered to be the standard formulation of SVM is due to Cortes and Vapnik (1995), who introduced the soft margin. A comprehensive textbook presentation of the SVM algorithm is provided by Vapnik (2000).

The kernel trick applies to many classic classifiers besides SVM, so we will discuss some of the theory in general. First, let us illustrate the idea with an example.

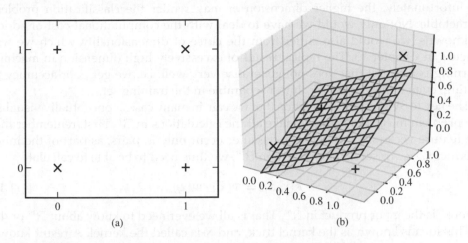

Figure 11.4 Feature space for the XOR problem: (a) in $\mathbb{R}^2$; (b) in $\mathbb{R}^3$

11.2.1 Example: The XOR Function

A standard example of linearly non-separable problems is the XOR function. We have two features $x_1, x_2 \in \{0, 1\}$ and two classes with labels given as $y = x_1 \oplus x_2 = x_1 + x_2 \mod 2$. This is illustrated in Figure 11.4(a). Clearly, no straight line can separate the two classes.

It is possible to separate the two classes by drawing a curve, but this would be a considerably more complex task, with more coefficients to optimise. Another alternative, which may not be intuitively obvious, is to add new features and transfer the problem into a higher-dimensional space where the classes become linearly separable.

In the XOR example, we could add a third feature $x_3 = x_1 x_2$. It is quite obvious that the four possible points (x_1, x_2, x_3) are not in the same plane and hence they span $\mathbb{R}^3$. Consequently, a plane can be drawn to separate any two points from the other two. In particular, we can draw a plane to separate the pluses and the crosses, as shown in Figure 11.4(b).

11.2.2 The SVM Algorithm

The XOR example illustrates that some problems which are not linearly separable may become so by introducing additional features as functions of the existing ones. In the example, we introduced a mapping $\Phi : \mathbb{R}^2 \to R^3$ from a two-dimensional feature space to a three-dimensional one, by the rule that

$$\Phi((x, y)) = (x, y, xy).$$

In general, we seek a mapping $\Phi : \mathbb{R}^n \to \mathbb{R}^m$ where $m > n$, so that the classification problem can be solved linearly in $\mathbb{R}^m$.

Unfortunately, the higher dimension m may render the classification problem intractable. Not only would we have to deal with the computational cost of added unknowns, we would also suffer from the curse of dimensionality which we will discuss in Chapter 13. The typical result of excessively high dimension in machine learning is a model that does not generalise very well, i.e. we get poor accuracy in testing, even if the classes are almost separable in the training set.

The key to SVM is the observation that we can, in many cases, conceptually visualise the problem in $\mathbb{R}^m$ but actually make all the calculations in $\mathbb{R}^n$. First, remember that the feature vectors $\mathbf{x}_i$ in the linear classifier occur only in pairs, as part of the inner product $\mathbf{x}_i \cdot \mathbf{x}_j$. To solve the problem in $\mathbb{R}^m$, we thus need to be able to calculate

$$K(\mathbf{x}_i, \mathbf{x}_j) = \Phi(\mathbf{x}_i) \cdot \Phi(\mathbf{x}_j), \tag{11.16}$$

where $\cdot$ is the inner product in $\mathbb{R}^m$. That is all we ever need to know about $\mathbb{R}^m$ or Φ.

This idea is known as the kernel trick, and K is called the kernel. A result known as Mercer's theorem states that if K satisfies certain properties, then there exists a Hilbert space H with a mapping $\Phi : \mathbb{R}^n \to H$ with an inner product satisfying (11.16). The notion of a Hilbert space generalises Euclidean spaces, so $\mathbb{R}^m$ is just a special case of a Hilbert space, and any Hilbert space will do for the purpose of our classifier.

A kernel K, in this context, is any continuous function $K : \mathcal{F} \times \mathcal{F} \to \mathbb{R}$, where $\mathcal{F} \subset \mathbb{R}^n$ is the feature space and $K(\mathbf{x}_i, \mathbf{x}_j) = K(\mathbf{x}_j, \mathbf{x}_i)$ (symmetric). To apply Mercer's theorem, the kernel also has to be non-negative definite (Mercer's condition), that is

$$\int \int K(\mathbf{x}_i, \mathbf{x}_j) g(\mathbf{x}) g(\mathbf{y}) d\mathbf{x}_i d\mathbf{x}_j \geq 0,$$

for all $g(\mathbf{x})$ such that

$$\int g(\mathbf{x})^2 d\mathbf{x} < +\infty.$$

Note that we can use any kernel function satisfying Mercer's condition, without identifying the space H. The linear hyperplane defined by $\mathbf{w}' \cdot \mathbf{x}' + w_0 = 0$ in H corresponds to a non-linear hypersurface in $\mathbb{R}^n$ defined by

$$K(\mathbf{w}, \mathbf{x}) + w_0 = 0.$$

Thus, the kernel trick gives us a convenient way to calculate a non-linear classifier while using the theory we have already developed for linear classifiers.

The kernel SVM classifier, for a given kernel K, is given simply by replacing all the dot products in the linear SVM solution by the kernel function. Hence we get

$$h(\mathbf{x}) = \sum_{i=1}^{N} \lambda_i y_i K(\mathbf{x}_i, \mathbf{x}) + w_0,$$

where the λ_i are found by solving

$$\max_{\lambda} \sum_{i=1}^{N} \lambda_i - \frac{1}{2} \sum_{i,j} \lambda_i \lambda_j y_i y_j K(\mathbf{x}_i, \mathbf{x}_j),$$

Table 11.1 Common kernel functions in machine learning

linear	$K(\mathbf{x}_i, \mathbf{x}_j) = \mathbf{x}_i \cdot \mathbf{x}_j$	
polynomial	$K(\mathbf{x}_i, \mathbf{x}_j) = (\gamma \mathbf{x}_i \cdot \mathbf{x}_j + r)^d$	$\gamma \geq 0$
radial basis function	$K(\mathbf{x}_i, \mathbf{x}_j) = e^{-\gamma \|\mathbf{x}_i - \mathbf{x}_j\|^2}$	$\gamma \geq 0$
sigmoid	$K(\mathbf{x}_i, \mathbf{x}_j) = \tanh(\gamma \mathbf{x}_i \cdot \mathbf{x}_j + r)$	

$$\text{subject to} \quad \sum_{i=1}^{N} \lambda_i y_i = 0,$$

$$0 \leq \lambda_i \leq C, \quad i = 1, 2, \ldots, N.$$

In the kernel SVM classifier, we can see that the number of support vectors may be critical to performance. When testing an object, by calculating $h(\mathbf{x})$, the kernel function has to be evaluated once for each support vector, something which may be costly. It can also be shown that the error rate may increase with increasing number of support vectors (Theodoridis and Koutroumbas, 2009).

There are four kernels commonly used in machine learning, as listed in Table 11.1. The linear kernel is just the dot product, and hence it is effectively as if we use no kernel at all. The result is a linear classifier in the original feature space. The polynomial kernel of degree d gives a classification hypersurface of degree d in the original feature space. The popular radial basis function that we used as the default in libSVM leads to a Hilbert space H of infinite dimension.

Example 11.2.1 *Recall the XOR example, where we introduced a third feature $X_3 = X_1 X_2$. This is equivalent to using the polynomial kernel with $d = 2$, $r = 0$ and $\gamma = 1$; we just need a scaling factor so that $X_3 = \sqrt{2}X_1 X_2$ and observe that $x^2 = x$ for $x \in \{0, 1\}$. Thus Φ is given as*

$$\Phi : \mathbf{x} = (x_1, x_2) \mapsto \mathbf{y} = (x_1^2, \sqrt{2}x_1 x_2, x_2^2).$$

Now, observe that

$$\mathbf{y}' \cdot \mathbf{y} = x_1'^2 x_1^2 + 2x_1' x_1 x_2' x_2 + x_2'^2 x_2^2 = (x_1' x_1 + x_2' x_2)^2 = (\mathbf{x}' \cdot \mathbf{x})^2,$$

which is just the polynomial kernel.

11.3 ν-SVM

There is a strong relationship between the SVM parameter C and the margin M. If C is large, a few outliers in the training set will have a serious impact on the classifier, because the error ϵ_i is considered expensive. If C is smaller, the classifier thinks little of such outliers and may achieve a larger margin. Even though the relationship between C and M is evident, it is not transparent. It is difficult to calculate the exact relationship and tune C analytically to obtain the desired result.

To get more intuitive control parameters, Schölkopf *et al.* (2000) introduced an alternative phrasing of the SVM problem, known as ν-SVM, which we present below. Note that ν-SVM works on exactly the same principles as the classic C-SVM already described. In particular, Chang and Lin (2000) showed that for the right choices of ν and C, both C-SVM and ν-SVM give the same solution.

In C-SVM, we normalised $\mathbf{w}$ and w_0 so that the boundary hyperplanes were defined as

$$\mathbf{x} \cdot \mathbf{w} + w_0 = \pm 1.$$

In ν-SVM, however, we leave the margin as a free variable ρ to be optimised, and let the boundary hyperplanes be given as

$$\mathbf{x} \cdot \mathbf{w} + w_0 = \pm \rho.$$

Thus, the minimisation problem can be cast as

$$\min \frac{1}{2} \|\mathbf{w}\|^2 - \nu\rho + \frac{1}{N} \sum_i \epsilon_i, \tag{11.17}$$

$$\text{subject to} \quad y_i(\mathbf{w} \cdot \mathbf{x} + w_0) \geq \rho - \epsilon_i, \quad i = 1, 2, \ldots, N, \tag{11.18}$$

$$\epsilon_i \geq 0, \quad i = 1, 2, \ldots, N, \tag{11.19}$$

$$\rho \geq 0. \tag{11.20}$$

The new parameter ν is used to weight the margin ρ in the cost function; the larger ν is, the more we favour a large margin at the expense of errors. Otherwise, the minimisation problem is similar to what we saw for C-SVM.

We will not go into all the details of solving the minimisation problem and analysing all the Lagrange multipliers. Interested readers can get a more complete discussion from Theodoridis and Koutroumbas (2009). Here we will just give some of the interesting conclusions. The Wolfe dual can be simplified to get

$$\max_{\boldsymbol{\lambda}} \quad \frac{1}{2} \sum_{i,j} \lambda_i \lambda_j y_i y_j \mathbf{x}_i \cdot \mathbf{x}_j, \tag{11.21}$$

$$\text{subject to} \quad \sum_{i=1}^{N} \lambda_i y_i = 0, \tag{11.22}$$

$$0 \leq \lambda_i \leq \frac{1}{N}, \quad i = 1, 2, \ldots, N, \tag{11.23}$$

$$\sum_{i=1}^{N} \lambda_i \geq \nu. \tag{11.24}$$

Contrary to the C-SVM problem, the cost function is now homogeneous quadratic. It can be shown that for every error $\epsilon_i > 0$, we must have $\lambda_i = 1/N$. Furthermore, unless $\rho = 0$, we will get $\sum \lambda_i = \nu$. Hence, the number of errors is bounded by νN, and the error rate on the training set is at most ν. In a similar fashion one can also show that the number of support vectors is at least νN.

The parameter ν thus gives a direct indication of the number of support vectors and the error rate on the training set. Finally, note that (11.23) and (11.24) imply $\nu \leq 1$. We also require $\nu \geq 0$, lest the margin be minimised.

11.4 Multi-class Methods

So far, we have viewed steganalysis as a binary classification problem, where we have exactly two possible class labels: 'stego' or 'clean'. Classification problems can be defined for any number of classes, and in steganalysis this can be useful, for instance, to identify the stego-system used for embedding. A multi-class steganalyser could for instance aim to recognise class labels like 'clean', 'F5', 'Outguess 0.2', 'JSteg', 'YASS', etc.

The support-vector machine is an inherently binary classifier. It finds one hyper-surface which splits the feature space in half, and it cannot directly be extended to the multi-class case. However, there are several methods which can be used to extend an arbitrary binary classifier to a multi-class classifier system.

The most popular method, known as *one-against-all*, will solve an M-class problem by employing M binary classifiers where each one considers one class against all others. Let $C_1, C_2, \ldots, C_M$ denote the M classes. The ith constituent classifier is trained to distinguish between the two classes

$$C_i \quad \text{and} \quad \bigcup_{j \neq i} C_j,$$

and using the SVM notation it gives a classifier function $h_i(\mathbf{x})$. Let the labels be defined so that $h_i(\mathbf{x}) > 0$ predicts $\mathbf{x} \in C_i$ and $h_i(\mathbf{x}) < 0$ predicts $\mathbf{x} \notin C_i$.

To test an object $\mathbf{y}$, each classification function is evaluated at $\mathbf{y}$. If all give a correct classification, there will be one unique i so that $h_i(\mathbf{y}) > 0$, and we would conclude that $\mathbf{y} \in C_i$. To deal with the case where one or more binary classifiers may give an incorrect prediction, we note that a heuristic close to zero is more likely to be in error than one with a high absolute value. Thus the multi-class classifier rule will be to assume

$$\mathbf{y} \in C_i \quad \text{for} \quad i = \arg \max_i h_i(\mathbf{y}). \tag{11.25}$$

An alternative approach to multi-class classification is *one-against-one*, which has been attributed to Knerr *et al.* (1990) (Hsu and Lin, 2002). This approach also splits the problem into multiple binary classification problems, but each binary class now compares one class C_i against one other C_j while ignoring all other classes. This requires $\binom{M}{2}$ binary classifiers, one for each possible pair of classes. This approach was employed in steganalysis by Pevný and Fridrich (2008). Several classifier rules exist for one-against-one classifier systems. The one used by Pevný and Fridrich is a simple majority voting rule. For each class C_i we count the binary classifiers that predict C_i, and we return the class C_i with the highest count. If multiple classes share the maximum count, we return one of them at random.

A comparison of different multi-class systems based on the binary SVM classifier can be found in Hsu and Lin (2002). They conclude that the classification accuracy is similar for both one-against-one and one-against-all, but one-against-one is significantly faster to train. The libSVM package supports multi-class classification using one-against-one, simply because of the faster training process (Chang and Lin, 2011).

The faster training time in one-against-one is due to the size of the training set in the constituent classifiers. In one-against-all, the entire training set is required for every constituent decoder, while one-against-one only uses objects from two classes at a time. The time complexity of SVM training is roughly cubic in the number of training vectors, and the number of binary classifiers is only quadratic in M. Hence, the one-against-one will be significantly faster for large M.

An additional benefit of one-against-one is flexibility and extendability. If a new class has to be introduced with an existing trained classifier, one-against-all requires complete retraining of all the constituent classifiers, because each requires the complete training set which is extended with a new class. This is not necessary in one-against-one. Since each classifier only uses two classes, all the existing classifiers are retained unmodified. We simply need to add M new constituent classifiers, pairing the new class once with each of the old ones.

11.5 One-class Methods

The two- and multi-class methods which we have discussed so far all have to be trained with actual steganograms as well as clean images. Hence they will never give truly blind classifiers, but be tuned to whichever stego-systems we have available when we create the training set. Even though unknown algorithms may sometimes cause artifacts similar to known algorithms, and thus be detectable, this is not the case in general. The ideal blind steganalyser should be able to detect steganograms from arbitrary, unknown stego-systems, even ones which were previously unknown to the steganalyst. In this case, there is only one class which is known at the time of training, namely the clean images. Since we do not know all stego-systems, we cannot describe the complement of this class of clean images.

One-class classification methods provide a solution. They are designed to be trained on only a single class, such as clean images. They distinguish between normal objects which look like training images, and abnormal objects which do not fit the model. In the case of steganalysis, abnormal objects are assumed to be steganograms.

This leads to truly blind steganalysis; no knowledge about the steganographic embedding is used, neither in algorithm design nor in the training phase. Thus, at least in principle, we would expect a classifier which can detect arbitrary and unknown steganalytic systems.

There is a one-class variant of SVM based on ν-SVM, described by Schölkopf *et al.* (2001). Instead of aiming to fit a hyperplane between the two known classes in the training set, it attempts to fit a tight hypersphere around the whole training set. In classification, objects falling within the hypersphere are assumed to be normal or negative, and those outside are considered abnormal or positive.

11.5.1 The One-class SVM Solution

A hypersphere with radius R centred at a point $\mathbf{c}$ is defined by the equation

$$(\mathbf{x} - \mathbf{c})^2 \leq R^2.$$

To train the one-class SVM, we thus need to find the optimal $\mathbf{c}$ and optimal R for the classification hypersphere.

Just as in the two-class case, we will want to accept a number of errors, that is training vectors which fall at a distance ϵ_i outside the hypersphere. This may not be quite as obvious as it was in the non-separable case, where errors would exist regardless of the solution. In the one-class case there is no second class to separate from at the training stage, and choosing R large enough, everything fits within the hypersphere. In contrast, the second class will appear in testing, and we have no reason to expect the two classes to be separable. Accepting errors in training means that the odd outliers will not force an overly large radius that would give a high false negative rate.

The minimisation problem is similar to the one we have seen before. Instead of maximising the margin, we now aim to keep the hypersphere as small as possible to give a tight description of the training set and limit false negatives as much as possible. Thus R^2 takes the place of $\|\mathbf{w}\|^2$ in the minimisation problem. As before we also need to minimise the errors ϵ_i. This gives the following minimisation problem:

$$\min R^2 + \frac{1}{\nu N} \sum \epsilon_i,$$

$$\text{subject to } \left\| \Phi(\mathbf{x}_i) - \mathbf{c} \right\|^2 \leq R^2 + \epsilon_i,$$

$$\epsilon_i \geq 0, \ i = 1, 2, \ldots, N,$$

where Φ is the mapping into the higher-dimensional space induced by the chosen kernel K, and $\mathbf{c}$ is the centre of the hypersphere in the higher-dimensional space. Again, we will skip the details of the derivation, and simply state the major steps and conclusions. The dual problem can be written as

$$\min_{\boldsymbol{\lambda}} \sum_{i,j} \lambda_i \lambda_j K(\mathbf{x}_i, \mathbf{x}_j) - \sum_i \lambda_i K(\mathbf{x}_i, \mathbf{x}_i),$$

$$\text{subject to } 0 \le \lambda_i \le \frac{1}{\nu N},$$

$$\sum_i \lambda_i = 1,$$

and the solution, defined as the centre of the hypersphere, is

$$\mathbf{c} = \sum_i \lambda_i \Phi(\mathbf{x}_i).$$

The squared distance between a point $\Phi(\mathbf{x})$ and the hypersphere centre $\mathbf{c}$ is given by

$$\|\mathbf{c} - \mathbf{x}\|^2 = \mathbf{c} \cdot \mathbf{c} - 2\mathbf{x} \cdot \mathbf{c} + \mathbf{x} \cdot \mathbf{x} = \sum_{i,j} \lambda_i \lambda_j K(\mathbf{x}_i, \mathbf{x}_j) - 2 \sum_i \lambda_i K(\mathbf{x}_i, \mathbf{x}) + K(\mathbf{x}, \mathbf{x}).$$

A point is classified as a positive if this distance is greater than R^2 and a negative if it is smaller. Hence we write the classification function as

$$h(\mathbf{x}) = \sum_{i,j} \lambda_i \lambda_j K(\mathbf{x}_i, \mathbf{x}_j) - 2 \sum_i \lambda_i K(\mathbf{x}_i, \mathbf{x}) + K(\mathbf{x}, \mathbf{x}) - R^2.$$

11.5.2 Practical Problems

The one-class SVM introduces a new, practical problem compared to the two-class case. The performance depends heavily on the SVM parameter ν or C and on the kernel parameters γ. In the two-class case, we solved this problem using a grid search with cross-validation. In the multi-class case, we can perform separate grid searches to tune each of the constituent (binary) classifiers. With a one-class training set, there is no way to carry out a cross-validation of accuracy.

The problem is that cross-validation estimates the accuracy, and the grid search seeks the maximum accuracy. With only one class we can only check for false positives; there is no sample of true positives which could be used to check for false negatives. Hence the grid search would select the largest R in the search space, and it would say nothing about the detection rate when we start testing on the second class.

Any practical application has to deal with the problem of parameter tuning. If we want to keep the true one-class training phase, there may actually be no way to optimise C or ν, simply because we have no information about the overlap between the training class and other classes. One possibility is to use a system of two-stage training. This requires a one-class training set T_1 which is used to train the support vector machine for given parameters C or v. Then, we need a multi-class training set T_2 which can be used in lieu of cross-validation to test the classifier for each parameter choice. This allows us to make a grid search similar to the one in the binary case. Observe that we called T_2 a training set, and not a test set. The parameters v or C will depend on T_2 and hence the tests done on T_2 are not independent and do not give information about the performance on unknown data. Where T_1 is used to train the system with respect to the SVM hypersphere, T_2 is used to train it with respect to C or v.

Similar issues may exist with respect to the kernel parameters, such as γ in the RBF. Since the parameter changes the kernel and thus the higher-dimensional space used, we cannot immediately assume that the radius which we aim to minimise is comparable for varying γ. Obviously, we can use the grid search with testing on T_2 to optimise γ together with C or v.

Remark 11.1 *It is possible to construct a multi-class classifier system using one-class classifiers instead of binary ones. One would need M constituent classifiers, one for each classifier, in a 'one-against-none' scheme. This was proposed and tested by Yang et al. (2007), using one-class SVM. They conclude that this gives classifiers as good as or better than previous schemes, with a shorter training time. As a very recent work, this solution has not been as extensively tested in the literature as the other multi-class techniques, and we have not seen it used in steganalysis.*

11.5.3 Multiple Hyperspheres

There are only few, but all the more interesting examples of one-class classifiers for steganalysis in the literature. Lyu and Farid (2004) used one-class SVM with Farid-72, but demonstrated that the hypersphere was the wrong shape to cover the training set. Using the one-class SVM system as described, they had accuracies around 10–20%.

Lyu and Farid suggested instead to use multiple, possibly overlapping hyperspheres to describe the training set. By using more than one hypersphere, each hypersphere would become more compact, with a smaller radius. Many hyperspheres together can cover complex shapes approximately. With six hyperspheres they achieved accuracies up to 70%.

Clearly, the SVM framework in itself does not support a multi-hypersphere model. However, if we can partition the training set into a sensible number of subsets, or clusters, we can make a one-class SVM model for each such cluster. Such a partition can be created using a clustering algorithm, like the ones we will discover in Section 12.4. Lyu and Farid (2004) used k-means clustering.

In experiments, the one-class SVM classifier with multiple hyperspheres tended to outperform two-class classifiers for short messages, while two-class classifiers were better for longer messages. The false positive rate was particularly low for the one-class classifiers. It should be noted, however, that this approach has not been extensively tested in the literature, and the available data are insufficient to conclude.

11.6 Summary

Support vector machines have proved to give a flexible framework, and off-the-shelf software packages provide one-, two- and multi-class classification. The vast majority of steganalysis research has focused on two-class classifiers. There is not yet sufficient data available to say much about the performance and applicability of one- and multi-class methods in general, and this leaves an interesting open problem.

12

Other Classification Algorithms

The problem of classification can be approached in many different ways. The support vector machine essentially solves a geometric problem of separating points in feature space. Probability never entered into the design or training of SVM. Probabilities are only addressed when we evaluate the classifier by testing it on new, unseen objects.

Many of the early methods from statistical steganalysis are designed as hypothesis tests and follow the framework of sampling theory within statistics. A null hypothesis H_0 is formed in such a way that we have a known probability distribution for image data assuming that H_0 is true. For instance, 'image I is a clean image' would be a possible null hypothesis. This gives precise control of the false positive rate. The principle of sampling theory is that the experiment is defined before looking at the observations, and we control the risk of falsely rejecting H_0 (false positives). The drawback of this approach is that this p-value is completely unrelated to the probability that H_0 is true, and thus the test does not give us any quantitative information relating to the problem.

The other main school of statistics, along side sampling theory, is known as Bayesian statistics or Bayesian inference. In terms of classification, Bayesian inference will aim to determine the probability of a given image I belonging to a given class C *a posteriori*, that is after considering the observed image I.

It is debatable whether SVM and statistical methods are learning algorithms. Some authors restrict the concept of machine learning to iterative and often distributed methods, such as neural networks. Starting with a random model, these algorithms inspect one training object at a time, adjusting the model for every observation. Very often it will have to go through the training set multiple times for the model to converge. Such systems learn gradually over time, and in some cases they can continue to learn also after deployment. At least new training objects can be introduced without discarding previous learning. The SVM algorithm, in contrast,

Machine Learning in Image Steganalysis, First Edition. Hans Georg Schaathun.
© 2012 John Wiley & Sons, Ltd. Published 2012 by John Wiley & Sons, Ltd.

starts with no model and considers the entire training set at once to calculate a model. We did not find room to cover neural networks in this book, but interested readers can find accessible introductions in books such as, for instance, Negnevitsky (2005) or Luger (2008).

In this chapter, we will look at an assortment of classifier algorithms. Bayesian classifiers, developed directly from Bayesian statistics, fit well into the framework already considered, of supervised learning and classification. Regression analysis is also a fundamental statistical technique, but instead of predicting a discrete class label, it predicts a continuous dependent variable, such as the length of the embedded message. Finally, we will consider unsupervised learning, where the machine is trained without knowledge of the true class labels.

12.1 Bayesian Classifiers

We consider an observed image I, and two possible events C_S that I be a steganogram and C_C that I be a clean image. As before, we are able to observe some feature vector $\mathbf{F}(I) = (F_1(I), \ldots, F_N(I))$. The goal is to estimate the *a posteriori* probability $P(C \mid \mathbf{F}(I))$ for $C = C_C, C_S$. This can be written using Bayes' law, as

$$P(C \mid \mathbf{F}(I)) = \frac{P(C) \cdot P(\mathbf{F}(I) \mid C)}{P(\mathbf{F}(I))}, \quad \text{for} \quad C = C_S, C_C. \tag{12.1}$$

The *a priori* probability $P(C_S) = 1 - P(C_C)$ is the probability that I is a steganogram, as well as we can gauge it, before we have actually looked at I or calculated the features $\mathbf{F}(I)$. The *a posteriori* probability $P(C_S \mid \mathbf{F}(I))$, similarly, is the probability *after* we have looked at I. The final two elements, $P(\mathbf{F}(I) \mid C)$ and $P(\mathbf{F}(I))$, describe the probability distribution of the features over a single class and over all images respectively. Note that (12.1) is valid whether $P(\mathbf{F})$ is a PMF or a PDF.

The Bayesian classifier is as simple as estimating $P(C \mid I)$ using (12.1), and concluding that the image is a steganogram if $P(C_S \mid I) > P(C_C \mid I)$, or equivalently if

$$P(C_S)P(\mathbf{F}(I) \mid C_S) > P(C_C)P(\mathbf{F}(I) \mid C_C),$$

$$\text{or equivalently,} \quad P(\mathbf{F}(I), C_S) > P(\mathbf{F}(I), C_C). \tag{12.2}$$

This, however, is not necessarily simple. The probability distribution of $\mathbf{F}(I)$ must be estimated, and we will discuss some relevant methods below. The estimation of the *a priori* probability $P(C)$ is more controversial; it depends heavily on the application scenario and is likely to be subjective.

The explicit dependence on $P(C)$ is a good thing. It means that the classifier is tuned to the context. If we employ a steganalyser to monitor random images on the Internet, $P(C_S)$ must be expected to be very small, and very strong evidence from $P(\mathbf{F}(I) \mid C_S)$ is then required to conclude that the image is a steganogram. If the steganalyser is used where there is a particular reason to suspect a steganogram, $P(C_S)$ will be larger, and less evidence is required to conclude that I is more likely a steganogram than not.

On the contrary how do we know what $P(C_S)$ is for a random image? This will have to rely on intuition and guesswork. It is customary to assume a uniform distribution, $P(C_S) = 0.5$, if no information is available, but in steganalysis we will often find ourselves in the position where we know that $P(C_S) = 0.5$ is far off, but we have no inkling as to whether $P(C_S) = 10^{-5}$ or $P(C_S) = 10^{-10}$ is closer to the mark.

12.1.1 Classification Regions and Errors

The classification problem ultimately boils down to identifying disjoint decision regions $R_S, R_C \subset \mathbb{R}^N$, where we can assume that I is a steganogram for all $\mathbf{F}(I) \in R_S$ and a clean image for all $\mathbf{F}(I) \in R_C$. It is possible to allow points $\mathbf{F}(I) \notin R_S \cup R_C$, which would mean that both class labels are equally likely for image I and we refuse to make a prediction. For simplicity we assume that $R_S \cup R_C = \mathbb{R}^N$, so that a prediction will be made for any image I.

In Figure 12.1 we show an example plot of $P(\mathbf{F}(I), C_C)$ and $P(\mathbf{F}(I), C_S)$ for a one-dimensional feature vector. The decision regions R_S and R_C will be subsets of the x-axis. The Bayes decision rule would define R_S and R_C so that every point x in the feature space would be associated with the class corresponding to the higher curve. In the plot, that means R_C is the segment to the left of the vertical line and R_S is the segment to the right.

The shaded area represents the classification error probability $P_e = 1 - A$, or Bayes' error, which can be defined as the integral

$$P_e = \int_{F \in R_C} P_{\mathbf{F}, C_S} d\mathbf{F} + \int_{F \in R_S} P_{\mathbf{F}, C_C} d\mathbf{F}. \qquad (12.3)$$

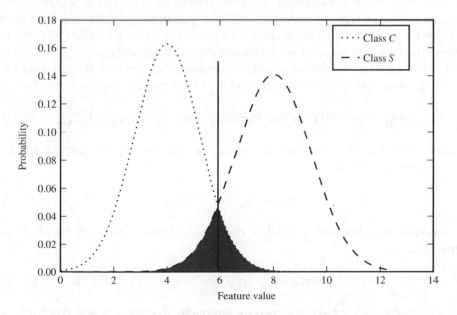

Figure 12.1 Feature probability distribution for different classes

If we move the boundary between the decision regions to the right, we will get fewer false positives, but more false negatives. Similarly, if we move it to the left we get fewer false negatives at the expense of false positives. Either way, the total error probability will increase. One of the beauties of the Bayes classifier is its optimality, in the sense that it minimises the probability P_e of classification error. This is not too hard to prove, as we shall see below.

Theorem 12.1.1 *The Bayes classification rule minimises the probability P_e of classification error.*

Proof. The decision rule in (12.2) defines two decision regions R_S and R_C, so that $\mathbf{F}(I) \in R_S$ when (12.2) is satisfied and $\mathbf{F}(I) \in R_C$ otherwise. To minimise the P_e expression in (12.3), R_C and R_S must be chosen such that every point $\mathbf{F}(I) \in \mathbb{R}^N$ appears in the integral with the smaller integrand, that is

$$R_S = \left\{ \mathbf{x} : P(C_C \mid \mathbf{x}) < P(C_S \mid \mathbf{x}) \right\},$$
$$R_C = \left\{ \mathbf{x} : P(C_S \mid \mathbf{x}) < P(C_C \mid \mathbf{x}) \right\},$$

which is exactly the Bayes decision rule.

12.1.2 Misclassification Risk

Risk captures both the probability of a bad event and the degree of badness. When different types of misclassification carry different costs, risk can be a useful concept to capture all error types in one measure. False positives and false negatives are two different bad events which often carry different costs. We are familiar with the error probabilities p_{FN} and p_{FP}, and we introduce the costs c_{FN} and c_{FP}.

The risk associated with an event is commonly defined as the product of the probability and cost. In particular, the risk of a false positive can be defined as

$$r_{FP} = c_{FP} \cdot P(FP) = c_{FP} \cdot P(FP|C_C) \cdot P(C_C) = c_{FP} \cdot p_{FP} \cdot P(C_C), \qquad (12.4)$$

where C_C denotes the event that we intercept a clean image. Similarly, the risk of a false negative would be

$$r_{FN} = c_{FN} \cdot p_{FN} \cdot P(C_S). \qquad (12.5)$$

The total risk of a classifier naturally is the sum of the risks associated with different events, i.e.

$$\text{risk} = r_{FP} + r_{FN} = c_{FP} \cdot p_{FP} \cdot P(C_C) + c_{FN} \cdot p_{FN} \cdot P(C_S). \qquad (12.6)$$

In this formulation, we have assumed for simplicity that there is zero cost (or benefit) associated with correct classifications. It is often useful to add costs or benefits c_{CP}

and c_{CN} in the event of correct positive and negative classification, respectively. Often c_{CP} and c_{CN} will be negative.

It is important to note that the cost does not have to be a monetary value, and there are at least two reasons why it might not be. Sometimes non-monetary cost measures are used because it is too difficult to assign a monetary value. Other times one may need to avoid a monetary measure because it does not accurately model the utility or benefit perceived by the user. Utility is an economic concept used to measure subjective value to an agent, that is Wendy in our case. One important observation is that utility does not always scale linearly in money. Quite commonly, individuals will think that losing £1000 is more than a hundred times worse than losing £10. Such a sub-linear relationship makes the user risk averse, quite happy to accept an increased probability of losing a small amount if it also reduces the risk of major loss.

In order to use risk as an evaluation heuristic for classifiers, we need to estimate c_{FP} and c_{FN} so that they measure the badness experienced by Wendy in the event of misclassification. The misclassification risk can be written as

$$r = c_{FP} \cdot \int_{R_S} P(\mathbf{F}(I), C_C) d\mathbf{F}(I) + c_{FN} \cdot \int_{R_C} P(\mathbf{F}(I), C_S) d\mathbf{F}(I),$$

assuming zero cost for correct classification.

We can define an *a posteriori* risk as well:

$$r^*(C_C) = c_{FN} \cdot P(C_C \mid \mathbf{F}(I)),$$
$$r^*(C_S) = c_{FP} \cdot (1 - P(C_S \mid \mathbf{F}(I))).$$

It is straightforward to rework the Bayesian decision rule to minimise the *a posteriori* risk instead of the error probability.

12.1.3 The Naïve Bayes Classifier

The Bayes classifier depends on the probability distribution of feature vectors. For one-dimensional feature vectors, this is not much of a problem. Although density estimation can be generalised for arbitrary dimension in theory, it is rarely practical to do so. The problem is that accurate estimation in high dimensions requires a very large sample, and the requirement grows exponentially.

The naïve Bayes classifier is a way around this problem, by assuming that the features are statistically independent. In other words, we assume that

$$P(\mathbf{F} \mid C) = \prod_{i=1}^{N} P(F_i \mid C), \quad \text{for } C = C_C, C_S. \tag{12.7}$$

This allows us to simplify (12.1) to

$$P(C \mid I) = \frac{P(C) \cdot \prod_{i=1}^{N} P(F_i(I) \mid C)}{\prod_{i=1}^{N} P(F_i(I))}. \tag{12.8}$$

Thus, we can estimate the probability distribution of each feature separately, using one-dimensional estimation techniques.

Of course, there is a reason that the classifier is called *naïve* Bayes. The independence assumption (12.7) will very often be false. However, the classifier has proved very robust to violations of the independence assumption, and in practice, it has proved very effective on many real-world data sets reported in the pattern recognition literature.

12.1.4 A Security Criterion

The Bayes classifier illustrates the fundamental criterion for classification to be possible. If the probability distribution of the feature vector is the same for both classes, that is, if

$$P(\mathbf{F}(I) \mid C_C) = P(\mathbf{F}(I) \mid C_S), \tag{12.9}$$

the Bayes decision rule will simply select the *a priori* most probable class, and predict the same class label regardless of the input. If there is no class skew, the output will be arbitrary.

The difference between the two conditional probabilities in (12.9) is a measure of security; the greater the difference, the more detectable is the steganographic embedding. A common measure of similarity of probability distributions is the relative entropy.

Definition 12.1.2 *The relative entropy (aka. Kullback–Leibler divergence) between two (discrete) probability distributions P_{Q_0} and P_{Q_1} is defined as*

$$D(P_{Q_0} \| P_{Q_1}) = \sum_{q \in \{x \in Q : P_{Q_0}(q) > 0\}} P_{Q_0}(q) \log \frac{P_{Q_0}(q)}{P_{Q_1}(q)}.$$

Note that the definition is only valid if $P_{Q_1}(q) > 0$ for all q such that $P_{Q_0}(q) > 0$.

We have $D(P_{Q_0} \| P_{Q_1}) \geq 0$, with equality if and only if the two distributions are equal. The relative entropy is often thought of as a distance, but it is not a distance (or metric) in the mathematical sense. This can easily be seen by noting that $D(P_1 \| P_2)$ is not in general equal to $D(P_2 \| P_1)$.

Cachin (1998) suggested using the relative entropy to quantify the security of a stego-system. Since security is an absolute property, it should not depend on any particular feature vector, and we have to consider the probability distribution of clean images and steganograms.

Definition 12.1.3 (Cachin, 1998) *Consider a stego-system X and let C be a cover and S a steganogram. Both C and S are considered as stochastic variables drawn from some set of* *possible images. We say that X is ε-secure against a passive warden if*

$$D(P_C||P_S) \leq \epsilon.$$

If ε = 0, the stego-system is perfectly secure.

Theorem 12.1.4 *If a stego-system is ε-secure, any steganalytic attack with false positive probability p_{FP} and false negative probability p_{FN} will satisfy*

$$\epsilon \geq p_{FP} \log \frac{p_{FP}}{1 - p_{FN}} + p_{FN} \log \frac{1 - p_{FP}}{p_{FN}}. \tag{12.10}$$

Proof. The steganalysis attack can be considered as a mapping of one random variable, namely the image I, into a Boolean stochastic variable which is either 'stego' or 'clean'. A standard result in information theory says that such deterministic processing cannot increase the relative entropy. Hence

$$D(B_C||B_S) \leq D(P_C||P_S) \leq \epsilon,$$

where B_C and B_S are the probability distributions of the predicted class label for clean images and steganograms respectively.

The right-hand side of (12.10) is just the binary relative entropy, sometimes denoted $d(p_{FP}, p_{FN}) = D(B_C||B_S)$, where B_C and B_S are binary stochastic variables with probability distributions $(p_{FP}, 1 - p_{FP})$ and $(p_{FN}, 1 - p_{FN})$ respectively.

Cachin's security criterion high lights a core point in steganography and steganalysis. Detectability is directly related to the difference between the probability distributions of clean images and steganograms. This difference can be measured in different ways; as an alternative to the Kullback–Leibler divergence, Filler and Fridrich (2009a,b) used Fisher information. In theory, Alice's problem can be described as the design of a cipher or stego-system which mimics the distribution of clean images, and Wendy's problem is elementary hypothesis testing.

In practice, the information theoretic research has hardly led to better steganalysis nor steganography. The problem is that we are currently not able to model the probability distributions of images. If Alice is not able to model P_C, she cannot design an ε-secure stego-system. In contrast, if Wendy cannot model P_C either, Alice probably does not need a theoretically ε-secure stego-system.

12.2 Estimating Probability Distributions

There are at least three major approaches to estimating probability density functions based on empirical data. Very often, we can make reasonable assumptions about the shape of the PDF, and we only need to estimate parameters. For instance, we might assume the distribution is normal, in which case we will only have to estimate the mean μ and variance σ^2 in order to get an estimated probability distribution. This is easily done using the sample mean $\bar{x}$ and sample variance s^2.

If we do not have any assumed model for the PDF, there are two approaches to estimating the distributions. The simplest approach is to discretise the distribution by dividing $\mathcal{X}$ into a finite number of disjoint bins $B_1, \ldots, B_m$ and use the histogram. The alternative is to use a non-parametric, continuous model. The two most well-known non-parametric methods are kernel density estimators, which we will discuss below, and k nearest neighbour.

12.2.1 The Histogram

Let X be a stochastic variable drawn from some probability distribution P_X on some continuous space $\mathcal{X}$. We can discretise the distribution by defining a discrete random variable X' as

$$X' = i \iff X \in B_i.$$

The probability mass function of X' is easily estimated by the histogram $h(i)$ of the observations of X'.

We can view the histogram as a density estimate, by assuming a PDF $\hat{f}$ which is constant within each bin:

$$\hat{f}(x) = \frac{1}{N} \cdot \frac{h(i)}{|B_i|}, \quad \text{for } x \in B_i,$$

where N is the number of observations and $|B|$ is the width (or volume) of the bin B. If the sample size N is sufficiently large, and the bin widths are wisely chosen, then $\hat{f}$ will indeed be a good approximation of the true PDF of X. The challenge is to select a good bin width. If the bins are too wide, any high-frequency variation in f will be lost. If the bins are too narrow, each bin will have very few observations, and $\hat{f}$ will be dominated by noise caused by the randomness of observations. In the extreme case, most bins will have zero or one element, with many empty bins. The result is that $\hat{f}$ is mainly flat, with a spike wherever there happens to be an observation.

There is no gospel to give us the optimal bin size, and it very often boils down to trial and error with some subjective evaluation of the alternatives. However, some popular rules of thumb exist. Most popular is probably Sturges' rule, which recommends to use $\hat{m} = 1 + \log_2 N$ bins, where N is the sample size. Interested readers can see Scott (2009) for an evaluation of this and other bin size rules.

12.2.2 The Kernel Density Estimator

Finally, we shall look at one non-parametric method, namely the popular kernel density estimator (KDE). The KDE is sometimes called Parzen window estimation, and the modern formulation is very close to the one proposed by Parzen (1962). The principles and much of the theory predates Parzen; see for instance Rosenblatt (1956), who is sometimes co-credited for the invention of KDE.

The fundamental idea in KDE is that every observation x_i is evidence of some density in a certain neighbourhood $\mathcal{N}(x_i)$ around x_i. Thus, x_i should make a positive contribution to $\hat{f}(x)$ for all $x \in \mathcal{N}(x_i)$ and a zero contribution for $x \notin \mathcal{N}(x_i)$. The KDE adds together the contributions of all the observations x_i.

The contribution made by each observation x_i can be modelled as a kernel function $K(x - x_i)$, where

$$K(x - x_i) = 0 \quad \text{for } x \notin \mathcal{N}(x_i),$$
$$K(x - x_i) > 0 \quad \text{for } x \in \mathcal{N}(x_i).$$

Summing over all the observations, the KDE estimate of the PDF is defined as

$$\hat{f}(x) = \frac{1}{N} \sum_{j=1}^{N} K(x - x_i). \qquad (12.11)$$

To be a valid PDF, $\hat{f}$ has to integrate to 1, and to ensure this, we require that the kernel also integrates to one, that is

$$\int_{-\infty}^{\infty} K(x)dx = 1.$$

It is also customary to require that the kernel is symmetric around 0, that is $K(x) = K(-x)$. It is easy to verify that if K is non-negative and integrates to 1, then so does $\hat{f}$ as defined in (12.11). Hence, $\hat{f}$ is a valid probability density function. Note that the kernel functions used here are not necessarily compatible with the kernels used for the kernel trick in Chapter 11, although they may sometimes be based on the same elementary functions, such as the Gaussian.

The simplest kernel is a hypercube, also known as a uniform kernel, defined as

$$K(x) = \begin{cases} 1 & \text{for } x_i < 0.5, \quad \forall i = 1, \dots, d, \\ 0 & \text{otherwise}, \end{cases}$$

where d is the dimensionality of x. The uniform kernel, which assigns a constant contribution to the PDF within the neighbourhood, will make a PDF looking like a staircase. To make a smoother PDF estimate, one would use a kernel function which is diminishing in the distance from the origin. Alongside the uniform kernel, the Gaussian kernel is the most commonly used. It is given by

$$K(x) = \frac{1}{(2\pi)^{d/2}} e^{\frac{1}{2}\|x\|^2},$$

and will give a much smoother PDF estimate.

The kernel density estimate is sensitive to scaling. This is easy to illustrate with the uniform kernel. If the sample space is scaled so that the variance of the observations x_i increases, then fewer observations will fall within the neighbourhood defined by the kernel. We would get the same effect as with a histogram with a very small bin width. Most bins or neighbourhoods are empty, and we get a short spike in $\hat{f}$ wherever there is an observation. In the opposite extreme, where the observations have very low variance, the PDF estimate would have absolutely no high-frequency information. In order to get a realistic estimate we thus need to scale either the kernel or the data in a way similar to tuning the bin width of a histogram.

The KDE analogue of the bin width is known as the *bandwidth*. The scaling can be incorporated into the kernel function, by introducing the so-called scaled kernel

$$K_h(\mathbf{x}) = \frac{1}{h^d} K\left(\frac{\mathbf{x}}{h}\right),$$

where the bandwidth h scales the argument to K. It is easily checked that if K integrates to 1, then so does K_h, thus K_h can replace K in the KDE definition (12.11). In the previous discussion, we related the scaling to the variance. It follows that the bandwidth should be proportional to the standard deviation so that the argument $\mathbf{x}/h$ to the kernel function has fixed variance.

The effect of changing bandwidth is illustrated in Figure 12.2. Scott's rule dictates a bandwidth of

$$h = N^{-\frac{1}{d+4}} s,$$

where s is the sample standard deviation. We can see that this gives a rather good estimate in the figure. If the bandwidth is reduced to a quarter, we see significant noise, and if it is four times wider, then the estimated PDF becomes far too flat. Code Example 12.1 shows how a KDE can be computed using the standard `scipy` library in Python. Only the Gaussian kernel is supported, and unfortunately, the bandwidth can only be changed by modifying the source code.

Code Example 12.1 Code example to plot the KDE estimate with Scott's bandwidth as shown in Figure 12.2

```
from scipy.stats.kde import gaussian_kde as gkde
import numpy as np
import matplotlib.pyplot as plt
S = np.random.randn( int(opt.size) )
X = [ i*0.1 for i in xrange(−40,40) ]
kde = gkde(S)    # Instantiate estimator
Y = kde(X)       # Evaluate estimated PDF
plt.plot( X, Y, "k−." )
```

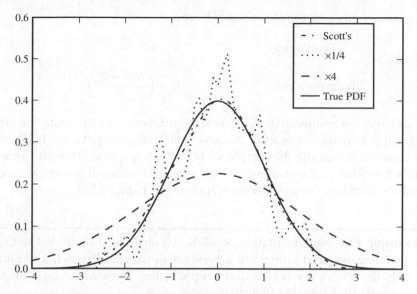

Figure 12.2 Examples of Gaussian KDE estimates from 100 pseudo-random points from a standard normal distribution

In the multivariate case $d > 1$, there is no reason to expect the same bandwidth to be optimal in all dimensions. To scale the different axes separately, we use a bandwidth matrix H instead of the scalar bandwidth h, and the scaled kernel can be written as

$$K_H(\mathbf{x}) = \frac{1}{|H|} K(H^{-1}\mathbf{x}), \tag{12.12}$$

where $|H|$ denotes the determinant of H. For instance, the scaled Gaussian kernel can be written as

$$K_H(\mathbf{x}) = \frac{1}{\sqrt{|2\pi H^2|}} \exp\left(\frac{1}{2}\mathbf{x}^T \cdot (H^{-1})^T \cdot H^{-1} \cdot \mathbf{x}\right).$$

In the scalar case, we argued that the bandwidth should be proportional to the standard deviation. The natural multivariate extension is to choose $H^T H = H^2$ proportional to the sample covariance matrix S^2, i.e.

$$H^2 = h^2 \cdot S^2,$$

for some scaling factor h. When we use the Gaussian kernel, it turns out that it suffices to calculate H^2; we do not need H directly. Because S is symmetric, we can rewrite the scaled Gaussian kernel as

$$K_H(\mathbf{x}) = \frac{1}{\sqrt{|2\pi H^2|}} \exp\left(\frac{1}{2}\mathbf{x}^T \cdot (H^2)^{-1} \cdot \mathbf{x}\right).$$

Two common choices for the scalar factor h are

$$\text{Scott's rule} \qquad h = N^{-\frac{1}{d+4}},$$

$$\text{Silverman's rule} \qquad h = \left(\frac{N(d+2)}{4}\right)^{-\frac{1}{d+4}}.$$

As we can see, the multivariate KDE is straightforward to calculate for arbitrary dimension d in principle. In practice, however, one can rarely get good PDF estimates in high dimension, because the sample set becomes too sparse. This phenomenon is known as the curse of dimensionality, and affects all areas of statistical modelling and machine learning. We will discuss it further in Chapter 13.

Code Example 12.2 Implementation of KDE. An object is instantiated with a data sample (training set) and optionally a bandwidth matrix. The default bandwidth matrix uses the covariance matrix and Scott's factor. The evaluate() method is used to evaluate the estimated PDF at arbitrary points

---

```python
from scipy import linalg
import numpy as np
class KDE(object):
    def __init__(self, dataset, bandwidth=None ):
        self.dataset = np.atleast_2d(dataset)
        self.d, self.n = self.dataset.shape
        self.setBandwidth( bandwidth )
    def evaluate(self, points):
        d, m = points.shape
        result = np.zeros((m,), points.dtype)
        for i in range(m):
            D = self.dataset - points[:,i,np.newaxis]
            T = np.dot( self.inv_bandwidth, D )
            E = np.sum( D*T, axis=0 ) / 2.0
            result[i] = np.sum( np.exp(-E), axis=0 )
        norm = np.sqrt(linalg.det(2*np.pi*self.bandwidth)) * self.n
        return result / norm
    def setBandwidth(self,bandwidth=None):
        if bandwidth != None: self.bandwidth = bandwidth
        else:
            factor = np.power(self.n, -1.0/(self.d+4)) # Scotts' factor
            covariance = np.atleast_2d(
                    np.cov(self.dataset, rowvar=1, bias=False) )
            self.bandwidth = covariance * factor * factor
        self.inv_bandwidth = linalg.inv(self.bandwidth)
```

An implementation of the KDE is shown in Code Example 12.2. Unlike the standard implementation in `scipy.stats`, this version allows us to specify a custom bandwidth matrix, a feature which will be useful in later chapters. The evaluation considers one evaluation point i at a time. Matrix algebra allows us to process every term of the sum in (12.11) with a single matrix operation. Thus, D and T have one column per term. The vector E, with one element per term, contains the exponents for the Gaussian and `norm` is the scalar factor $\sqrt{2\pi \left| H^2 \right|}$.

12.3 Multivariate Regression Analysis

Regression denotes the problem of predicting one stochastic variable Y, called the *response variable*, by means of observations of some other stochastic variable X, called the *predictor variable* or independent variable. If X is a vector variable, for instance a feature vector, we talk about multivariate regression analysis. Where classification aims to predict a class label from some discrete set, regression predicts a continuous variable $Y \in \mathbb{R}$.

Let $Y \in \mathbb{R}$ and $X \in \mathbb{R}^n$. The underlying assumption for regression is that Y is approximately determined by X, that is, there is some function $f : \mathbb{R}^n \to \mathbb{R}$ so that

$$Y = f(X) + \eta, \tag{12.13}$$

where η is some small, random variable. A regression method is designed to calculate a function $\hat{f} : \mathbb{R}^n \to \mathbb{R}$ which estimates f. Thus, $\hat{f}(\mathbf{x})$ should be an estimator for Y, given an observation $\mathbf{x}$ of X.

Multivariate regression can apply to steganalysis in different ways. Let the predictor variable X be a feature vector from some object which may be a steganogram. The response variable Y is some unobservable property of the object. If we let Y be the length of the embedded message, regression can be used for quantitative steganalysis. Pevný *et al.* (2009b) tested this approach using support vector regression. Regression can also be used for binary classification, by letting Y be a numeric class label, such as ± 1. Avcibaş *et al.* (2003) used this technique for basic steganalysis.

12.3.1 Linear Regression

The simplest regression methods use a linear regression function $\hat{f}(X)$, i.e.

$$\hat{f}(X_1, X_2, \ldots, X_m) = w_0 + \sum_{i=1}^{m} w_i X_i = \sum_{i=0}^{m} w_i X_i, \tag{12.14}$$

where we have fixed $X_0 \equiv 1$ for the sake of brevity. Obviously, we need to choose the w_is to get the best possible prediction, but what do we mean by best possible?

We have already seen linear regression in action, as it was used to calculate the prediction error image for Farid's features in Section 7.3.2. The assumption was that each wavelet coefficient Y can be predicted using a function $\hat{f}$ of coefficients $X = (X_1, X_2, \ldots, X_m)$, including adjacent coefficients, corresponding coefficients from adjacent wavelet components and corresponding coefficients from the other colour channels. Using the regression formula $Y = \hat{f}(X) + \eta$, Y denotes the image being tested, $\hat{f}(X)$ is the prediction image and η the prediction error image.

The most common criterion for optimising the regression function is the *method of least squares*. It aims to minimise the squared error η^2 in (12.13). The calculation of the regression function can be described as a training phase. Although the term is not common in statistics, the principle is the same as in machine learning. The input to the regression algorithm is a sample of observations $y_1, \ldots, yM$ of the response variable Y, and corresponding observations $x_1, \ldots, x_M$ of the predictor variable. The assumption is that y_i depends on x_i.

The linear predictor can be written in matrix form as

$$y = \hat{f}(\mathbf{x}) = \mathbf{w} \cdot \mathbf{x},$$

where the observation $x_0 = 1$ of the deterministic dummy variable X_0 is included in $\mathbf{x}$. Let $\mathbf{y} = (y_1, \ldots, y_M)^\mathrm{T}$ be a column vector containing all the observed response variables, and let $\mathbf{X}$ be an $M \times N$ matrix containing the observed feature vectors as rows. The prediction errors on the training set can be calculated simultaneously using a matrix equation:

$$\boldsymbol{\eta} = (\eta_1, \ldots, \eta_M)^\mathrm{T} = \mathbf{y} - \mathbf{w} \cdot \mathbf{X}.$$

Thus the least-squares regression problem boils down to solving the optimisation problem

$$\min_{\mathbf{w}} \|\boldsymbol{\eta}\|^2,$$

where

$$\|\boldsymbol{\eta}\|^2 = \|\mathbf{y} - \mathbf{Xw}\|^2 = \mathbf{y}^\mathrm{T}\mathbf{y} - \mathbf{y}^\mathrm{T}\mathbf{Xw} - \mathbf{w}^\mathrm{T}\mathbf{X}^\mathrm{T}\mathbf{y} + \mathbf{w}^\mathrm{T}\mathbf{X}^\mathrm{T}\mathbf{Xw}.$$

The minimum is the point where the derivative vanishes. Thus we need to solve

$$0 = \frac{\partial}{\partial \mathbf{w}} \|\mathbf{y} - \mathbf{Xw}\|^2 = -2\mathbf{y}^\mathrm{T}\mathbf{X} + 2\mathbf{w}^\mathrm{T}\mathbf{X}^\mathrm{T}\mathbf{X}.$$

Hence, we get the solution given as

$$\mathbf{w} = (\mathbf{X}^\mathrm{T}\mathbf{X})^{-1}\mathbf{X}^\mathrm{T}\mathbf{y},$$

and the regression function (12.14) reads

$$f(\mathbf{x}) = ((\mathbf{X}^\mathrm{T}\mathbf{X})^{-1}\mathbf{X}^\mathrm{T}\mathbf{y})^\mathrm{T}\mathbf{x}.$$

This is the same formula we used for the linear predictor of Farid's features in Chapter 7.

12.3.2 Support Vector Regression

There is also a regression algorithm based on support vector machines. Schölkopf *et al.* (2000) even argue that support vector regression (SVR) is simpler than the classification algorithm, and that it is the natural first stage.

Note that the usual SVM algorithm results in a function $h : \mathbb{R}^N \to \mathbb{R}$, which may look like a regression function. The only problem with the standard SVM is that it is not optimised for regression, only for classification. When we formulate support vector regression, the function will have the same structure:

$$h(\mathbf{x}) = \sum w_i K(\mathbf{x}, \mathbf{x_i}) + w_0, \tag{12.15}$$

but the w_i are tuned differently for classification and for regression.

A crucial property of SVM is that a relatively small number of support vectors are needed to describe the boundary hyperplanes. In order to carry this sparsity property over to regression, Vapnik (2000) introduced the ε-insensitive loss function:

$$\|y - f(\mathbf{x})\|_\varepsilon := \max\{0, |y - f(\mathbf{x})| - \varepsilon\}. \tag{12.16}$$

In other words, prediction errors η less than ε are accepted and ignored. The threshold $\varepsilon \geq 0$ is a free variable which can be optimised.

This leads to a new optimisation problem, which is similar to those we saw in Chapter 11:

$$\min \tau(\mathbf{w}, \boldsymbol{\xi}, \boldsymbol{\xi}^*, \varepsilon) = \frac{1}{2} \|\mathbf{w}\|^2 + C \left(v\varepsilon + \frac{1}{\ell} \sum_{i=1}^{\ell} (\xi_i + \xi_i^*) \right),$$

$$\text{subject to } (\mathbf{w} \cdot \mathbf{x}_i + b) - y_i \leq \varepsilon + \xi_i,$$

$$y_i - (\mathbf{w} \cdot \mathbf{x}_i + b) \leq \varepsilon + \xi_i^*,$$

$$\xi_i, \xi_i^*, \varepsilon \geq 0.$$

The constants v and C are design parameters which are used to control the trade-off between minimising the slack variables ξ_i and ξ_i^*, the model complexity $\|\mathbf{w}\|^2$ and the sensitivity threshold ε. The constraints simply say that the regression error $|\eta|$ should be bounded by ε plus some slack, and that the threshold and slack variables should be non-negative.

Using the standard techniques as with other SVM methods, we end up with a dual and simplified problem as follows:

$$\max W(\boldsymbol{\lambda}, \boldsymbol{\lambda}^*) = \sum_{i=1}^{\ell} (\lambda_i^* - \lambda_i) y_i - \frac{1}{2} \sum_{i=1}^{\ell} \sum_{j=1}^{\ell} (\lambda_i^* - \lambda_i)(\lambda_j^* - \lambda_j) K(\mathbf{x}_i, \mathbf{x}_j),$$

subject to

$$\sum_{i=1}^{\ell}(\lambda_i^* - \lambda_i) = 0,$$

$$0 \le \lambda_i, \lambda_i^* \le \frac{C}{\ell}, \quad \text{for } i = 1, \ldots, \ell,$$

$$\sum_{i=1}^{\ell}(\lambda_i^* + \lambda_i) \le Cv.$$

The solutions λ_i and λ_i^* are used to form the regression function

$$f(\mathbf{x}) = \sum_{i=1}^{\ell}(\lambda_i^* + \lambda_i)K(\mathbf{x}_i, \mathbf{x}) + b.$$

The most significant difference between this regression function and that from least-squares linear regression is of course that the kernel trick makes it non-linear. Hence much more complex dependencies can be modelled. There is support for SVM regression in libSVM.

12.4 Unsupervised Learning

The classifiers we have discussed so far employ supervised learning. That is, the input to the training algorithm is feature vectors with class labels, and the model is fitted to the known classes. In unsupervised learning, no labels are available to the training algorithm. When we use unsupervised learning for classification, the system will have to identify its own classes, or clusters, of objects which are similar. This is known as cluster analysis or clustering. Similarity often refers to Euclidean distance in feature space, but other similarity criteria could be used. This may be useful when no classification is known, and we want to see if there are interesting patterns to be recognised.

In general, supervised learning can be defined as learning in the presence of some error or reward function. The training algorithm receives feedback based on what the correct solution should be, and it adjusts the learning accordingly. In unsupervised learning, there is no such feedback, and the learning is a function of the data alone.

The practical difference between a 'cluster' as defined in cluster analysis and a 'class' as considered in classification is that a class is strictly defined whereas a cluster is more loosely defined. For instance, an image is either clean or a steganogram, and even though we cannot necessarily tell the difference by inspecting the image, the class is strictly determined by how the image was created. Clusters are defined only through the appearance of the object, and very often there will be no objective criteria by which to determine cluster membership. Otherwise clustering can be thought of, and is often used, as a method of classification.

We will only give one example of a statistical clustering algorithm. Readers who are interested in more hard-core learning algorithms, should look into the self-organising maps of Kohonen (1990).

12.4.1 K-means Clustering

The K-means clustering algorithm is a standard textbook example of clustering, and it is fairly easy to understand and implement. The symbol K just denotes the number of clusters. A sample S of n objects, or feature vectors, $\mathbf{x}_i \in \mathbb{R}^N$ is partitioned into K clusters $C_1, C_2, \ldots, C_K$.

Each cluster C_j is identified by its mean $\mathbf{m}_j$, and each object $\mathbf{x}_i$ is assigned to the cluster corresponding to the closest mean, that is to minimise $\|\mathbf{x}_i - \mathbf{m}_j\|$. The solution is found iteratively. When the mean is calculated, the cluster assignments have to change to keep the distances minimised, and when the cluster assignment changes, the means will suddenly be wrong and require updating. Thus we alternate between updating the means and the cluster assignments until the solution converges. To summarise, we get the following algorithm:

1. Choose K random points $\mathbf{m}_j$ for $j = 1, \ldots, K$.
2. Form K disjoint sets $C_j \subset S$, so that $\mathbf{x}_i \in C_j$ if j solves $\min_j \|\mathbf{x}_i - \mathbf{m}_j\|$.
3. Update $\mathbf{m}_j$ to be the sample mean of C_j, i.e.

$$\mathbf{m}_j = \frac{1}{\#C_j} \sum_{\mathbf{x} \in C_j} \mathbf{x}.$$

4. Repeat from Step 2 until the solution converges.

It can be shown that the solution always converges.

Code Example 12.3 shows how this can be done in Python. The example is designed to take advantage of efficient matrix algebra in numpy, at the expense of readability.

Code Example 12.3 K-means clustering in Python; the input X is an $N \times m$ array with m feature vectors as rows. The output is a pair (C, M), where C is an integer array with element i the index of the cluster containing $\mathbf{x}_i$ and M is a $N \times K$ array where the ith row is the mean of cluster i.

```python
def kmeans(X,K=2,eps=0.1):
  (N,m) = X.shape
  M = np.random.randn(N,m)
  while True:
    L = [ np.sum( (X − M[[i],:])**2, axis=1 ) for i in xrange(K) ]
    A = np.vstack( L )
    C = np.argmin( A, axis=0 )
    L = [ np.mean( X[(C==i),:], axis=0 ) for i in xrange(K) ]
    M1 = np.vstack(M1)
    if np.sum((M−M1)**2) < eps: return (C,M1)
    else: M = M1
```

The rows of the input X are the feature vectors. Each row of M is the mean of one cluster, and is initialised randomly. The matrix A is $K \times m$, where $A_{j,i} = \|x_i - m_j\|^2$. Then C is calculated so that the ith element is the index of the cluster containing x_i, which corresponds to the row index of the maximal element in each column of A. Finally, M_1 is calculated as the new sample mean of each cluster, to replace M. If the squared Euclidean distance between M and M_1 is smaller than a preset threshold we return, otherwise we reiterate.

Criticism and Drawbacks

The K-means algorithm is very intuitive and relatively fast. One may say it is too simplistic in that each cluster is defined only by its sample mean, without any notion of shape, density or volume. This can have some odd effects when the cluster shapes are far from spherical. Figure 12.3, for example, shows two elongated clusters, and human observers would probably focus on the gap between the upper and lower clusters. The K-means algorithm, implicitly looking for spherical clusters, divides it into a lefts-hand and a right-hand cluster each containing about half each of the elongated clusters. Similar problems can occur if the clusters have very different sizes.

Another question concerns the distance function, which is somewhat arbitrary. There is no objective reason to choose any one distance function over another. Euclidean distance is common, because it is the most well-known distance measure. Other distance functions may give very different solutions. For example, we can

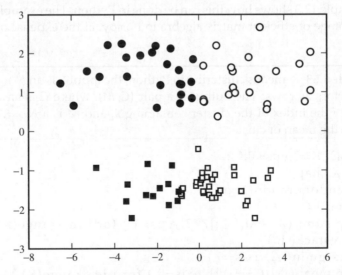

Figure 12.3 Elongated clusters with K-means. Two sets, indicated by squares and circles, were generated using a Gaussian distribution with different variances and means. The colours, black and white, indicate the clustering found by K-means for $K = 2$

apply the kernel trick, and calculate the distance as

$$d(\mathbf{x}, \mathbf{m})^2 = K(\mathbf{x} - \mathbf{m}, \mathbf{x} - \mathbf{m})^2, \tag{12.17}$$

for some kernel function K.

12.5 Summary

The fundamental classification problem on which we focused until the start of this chapter forms part of a wider discipline of pattern recognition. The problem has been studied in statistics since long before computerised learning became possible. Statistical methods, such as Bayesian classifiers in this chapter and Fisher linear discriminants in the previous one, tend to be faster than more modern algorithms like SVM.

We have seen examples of the wider area of pattern recognition, in unsupervised learning and in regression. Regression is known from applications in quantitative steganalysis, and unsupervised learning was used as an auxiliary function as part of the one-class steganalyser with multiple hyperspheres of Lyu and Farid (2004) (see Section 11.5.3). Still, we have just scratched the surface.

13

Feature Selection
and Evaluation

Selecting the right features for classification is a major task in all areas of pattern matching and machine learning. This is a very difficult problem. In practice, adding a new feature to an existing feature vector may increase or decrease performance depending on the features already present. The search for the perfect vector is an NP-complete problem. In this chapter, we will discuss some common techniques that can be adopted with relative ease.

13.1 Overfitting and Underfitting

In order to get optimal classification accuracy, the model must have just the right level of complexity. Model complexity is determined by many factors, one of which is the dimensionality of the feature space. The more features we use, the more degrees of freedom we have to fit the model, and the more complex it becomes.

To understand what happens when a model is too complex or too simple, it is useful to study both the training error rate ϵ^* and the testing error rate ϵ. We are familiar with the testing error rate from Chapter 10, where we defined the accuracy as $A = 1 - \epsilon$, while the training error rate ϵ^* is obtained by testing the classifier on the training set. Obviously, it is only the testing error rate which gives us any information about the performance on unknown data.

If we start with a simple model with few features and then extend it by adding new features, we will typically see that ϵ^* decreases. The more features, the better we can fit the model to the training data, and given enough features and thus degrees of freedom, we can fit it perfectly to get $\epsilon^* = 0$. This is illustrated in Figure 13.1, where we see the monotonically decreasing training error rate.

Now consider ϵ in the same way. In the beginning we would tend to have $\epsilon \approx \epsilon^*$, where both decrease with increasing complexity. This situation is known as

Machine Learning in Image Steganalysis, First Edition. Hans Georg Schaathun.
© 2012 John Wiley & Sons, Ltd. Published 2012 by John Wiley & Sons, Ltd.

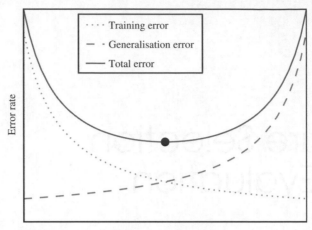

Feature space dimension

Figure 13.1 Training error rate and testing error rate for varying dimensionality (for illustration)

underfitting. The model is too simple to fit the data very well, and therefore we get poor classification on the training and test set alike. After a certain point, ϵ will tend to rise while ϵ^* continues to decline. It is natural that ϵ^* declines as long as degrees of freedom are added. When ϵ starts to increase, so that $\epsilon \gg \epsilon^*$, it means that the model no longer generalises for unknown data. This situation is known as overfitting, where the model has become too complex. Overfitting can be seen as support for Occam's Razor, the principle that, other things being equal, the simpler theory is to be preferred over the more complex one. An overfitted solution is inferior because of its complexity.

It may be counter-intuitive that additional features can sometimes give a decrease in performance. After all, more features means more information, and thus, in theory, the classifier should be able to make a better decision. This phenomenon is known as the *curse of dimensionality*. High-dimensionality feature vectors mean that the classifier gets many degrees of freedom and the model can be fitted perfectly to the training data. Thus, the model will not only capture statistical properties of the different classes, but also random noise elements which are peculiar for each individual object. Such a model generalises very poorly for unseen objects, leading to the discrepancy $\epsilon \gg \epsilon^*$. Another way to view the curse of dimensionality is to observe that the distance between points in space increases as the dimension of the space increases. Thus the observations we make become more scattered and it seems reasonable that less can be learnt from them.

There are other causes of overfitting, besides the curse of dimensionality. In the case of SVM, for instance, tuning the parameter C (or ν) is critical, as it controls the weighting of the two goals of minimising the (training) errors and maximising the margin. If C is too large, the algorithm will prefer a very narrow margin to avoid errors, and a few outliers which are atypical for the distribution can have an undue impact on the classifier. This would push ϵ up and ϵ^* down, indicating another example of overfitting. Contrarily, if C is too small, the importance of errors is discounted, and the classifier will accept a fairly high ϵ^* giving evidence of underfitting. Thus, the problems with under- and overfitting make up one of the motivations behind the grid search and cross-validation that we introduced in Section 3.3.2.

13.1.1 Feature Selection and Feature Extraction

In order to avoid the curse of dimensionality, it is often necessary to reduce the dimension of the feature space. There are essentially two approaches to this, namely feature selection and feature extraction.

Feature extraction is the most general approach, aiming to map feature vectors in $\mathbb{R}^n$ into some smaller space $\mathbb{R}^m$ ($m < n$). Very often, almost all of the sample feature vectors are approximately contained in some m-dimensional subspace of $\mathbb{R}^n$. If this is the case, we can project all the feature vectors into $\mathbb{R}^m$ without losing any structure, except for the odd outlier. Even though the mapping $d : \mathbb{R}^n \to \mathbb{R}^m$ may in general sacrifice information, when the sample feature vectors really span a space of dimension greater than m, the resulting feature vectors in $\mathbb{R}^m$ may still give superior classification by avoiding overfitting. There are many available techniques for feature extraction, with principal component analysis (PCA) being one of the most common examples. In steganalysis, Xuan *et al.* (2005b) used a feature extraction method of Guorong *et al.* (1996), based on the Bhattacharyya distance. The details are beyond the scope of this book, and we will focus on feature selection in the sequel.

Feature selection is a special and limited case of feature extraction. Writing a feature vector as $\mathbf{x} = (x_i \mid i = 1, \ldots, n)$, we are only considering maps $d : \mathbb{R}^n \to \mathbb{R}^m$ which can be written as

$$d : (x_i \mid i = 1, \ldots, n) \mapsto (x_i \mid i \in \mathcal{D}),$$

for some subset $\mathcal{D} \subset \{1, 2, \ldots, n\}$. In other words, we select individual features from the feature vector $\mathbf{x}$, ignoring the others completely.

In addition to reducing the dimensionality to avoid overfitting, feature selection is also a good help to interpret the problem. A classifier taking a 1000-D feature vector to get a negligible error rate may be great to solve the steganalyser's practical problem. However, it is almost impossible to analyse, and although the machine can learn a lot, the user learns rather little.

If we can identify a low-dimensional feature vector with decent classification performance, we can study each feature and find out how and why it is affected by the embedding. Thus a manageable number of critical artifacts can be identified, and the design of new steganographic systems can focus on eliminating these artifacts. Using feature selection in this context, we may very well look for a feature vector giving inferior classification accuracy. The goal is to find one which is small enough to be manageable and yet explains most of the detectability. Because each feature is either used or ignored, feature selection gives easy-to-use information about the artifacts detected by the classifier.

13.2 Scalar Feature Selection

Scalar feature selection aims to evaluate individual, scalar features independently. Although such an approach disregards dependencies between the different features in a feature vector, it still has its uses. An important benefit is that the process scales linearly in the number of candidate features. There are many heuristics and approaches which can be used. We will only have room for ANOVA, which was used to design one of the pioneering feature vectors in steganalysis.

13.2.1 Analysis of Variance

Analysis of variance (ANOVA) is a technique for hypothesis testing, aiming to determine if a series of sub-populations have the same mean. In other words, if a population is made up of a number of disjoint classes $C_1, C_2, \ldots, C_N$ and we observe a statistic X, we want to test the null hypothesis:

$$H_0 : \mu_1 = \mu_2 = \ldots = \mu_N,$$

where μ_i is the mean of X when drawn from C_i. A basic requirement for any feature in machine learning is that it differs between the different classes C_i. Comparing the means of each class is a reasonable criterion.

Avcibaş et al. (2003) started with a large set of features which they had studied in a different context and tested them on a steganalytic problem. They selected the ten features with the lowest p-value in the hypothesis test to form their IQM feature vector (see Section 6.2). It is true that a feature may be discarded because it has the same mean for all classes and still have some discriminatory power. However, the ANOVA test does guarantee that those features which are selected will have discriminatory power.

There are many different ANOVA designs, and entire books are devoted to the topic. We will give a short presentation of one ANOVA test, following Johnson and Wichern (1988), but using the terminology of steganalysis and classification. We have a feature X representing some property of an object O which is drawn randomly from a population consisting of a number of disjoint classes C_ℓ, for $\ell = 1, \ldots, N$. Thus X is a stochastic variable. We have a number of observations $X_{\ell,i}$, where $i = 1, 2, \ldots, n_\ell$ and $X_{\ell,i}$ is drawn from C_ℓ.

There are three assumptions underlying ANOVA:

1. The observations $X_{\ell,i}$ are independent.
2. All the classes have the same variance σ_ℓ^2.
3. Each class is normally distributed.

The first assumption should be no surprise. We depend on this whenever we test a classifier to estimate error probabilities too, and it amounts to selecting media objects randomly and independently. The third assumption, on the distribution, can be relaxed if the sample is sufficiently large, due to the Central Limit Theorem.

The second assumption, that the variances are equal, gives cause for concern. *A priori*, there is no reason to think that steganographic embedding should not have an effect on the variance instead of or in addition to the mean. Hence, one will need to validate this assumption if the quantitative conclusions made from the ANOVA test are important.

To understand the procedure, it is useful to decompose the statistic or feature into different effects. Each observation can be written as a sum

$$X_{\ell,j} = \mu + \tau_\ell + e_{\ell,j},$$

where μ is the mean which is common for the entire population and τ_ℓ is a so-called class effect which is the difference between the class mean and the population mean, i.e. $\tau_\ell = \mu_\ell - \mu$. The last term can be called a random error $e_{\ell,j}$, that is the effect due to the individuality of each observation.

With the notation of the decomposition, the null hypothesis may be rephrased as

$$H_0 : \tau_1 = \tau_2 = \ldots = \tau_N.$$

In practice the means are unknown, but we can make an analogous decomposition using sample means as follows:

$$x_{\ell,j} = \bar{x} + (\bar{x}_\ell - \bar{x}) + (x_{\ell,j} - \bar{x}_\ell), \tag{13.1}$$

for $\ell = 1, 2, \ldots, N$ and $j = 1, 2, \ldots, n_\ell$. This can be written as a vector equation,

$$\mathbf{x} = \mathbf{1}\bar{x} + \mathbf{t} + \mathbf{e}, \tag{13.2}$$

where $\mathbf{1}$ is an all-one vector, and

$$\mathbf{x} = (x_{1,1}, \ldots, x_{1,n_1}, x_{2,1}, \ldots, x_{2,n_1}, x_{3,n_3}, \ldots, x_{N-1,n_{N-1}}, x_{N,n_N}, \ldots, x_{N,n_N}),$$
$$\mathbf{t} = (\bar{x}_1 - \bar{x}, \ldots, \bar{x}_1 - \bar{x}, \bar{x}_2 - \bar{x}, \ldots, \bar{x}_{N-1} - \bar{x}, \bar{x}_N - \bar{x}, \ldots, 1, \bar{x}_N - \bar{x}),$$
$$\mathbf{e} = \mathbf{x} - \bar{x}\mathbf{1}.$$

We are interested in a decomposition of the sample variance $(\mathbf{x} - \mathbf{1}\bar{x})^2$, where squaring denotes an inner product of a vector and itself. We will show that

$$(\mathbf{x} - \mathbf{1}\bar{x})^2 = \mathbf{t}^2 + \mathbf{e}^2. \tag{13.3}$$

This equation follows from the fact that the three vectors on the right-hand side of (13.2) are orthogonal. In other words, $\mathbf{1} \cdot \mathbf{t} = 0$, $\mathbf{1} \cdot \mathbf{e} = 0$ and $\mathbf{t} \cdot \mathbf{e} = 0$ regardless of the actual observations. To demonstrate this, we will give a detailed derivation of (13.3). Subtracting $\bar{x}$ on both sides of (13.1) and squaring gives

$$(X_{\ell,j} - \bar{x})^2 = (\bar{x}_\ell - \bar{x})^2 + (x_{\ell,j} - \bar{x}_\ell)^2 + 2(\bar{x}_\ell - \bar{x})(x_{\ell,j} - \bar{x}_\ell),$$

and summing over j, while noting that $\sum(x_{\ell,j} - \bar{x}_\ell) = 0$, gives

$$\sum_{j=1}^{n_\ell}(X_{\ell,j} - \bar{x})^2 = n_\ell(\bar{x}_\ell - \bar{x})^2 + \sum_{j=1}^{n_\ell}(x_{\ell,j} - \bar{x}_\ell)^2.$$

Next, summing over ℓ, we get

$$\sum_{\ell=1}^{N}\sum_{j=1}^{n_\ell}(X_{\ell,j} - \bar{x})^2 = \sum_{\ell=1}^{N} n_\ell(\bar{x}_\ell - \bar{x})^2 + \sum_{\ell=1}^{N}\sum_{j=1}^{n_\ell}(x_{\ell,j} - \bar{x}_\ell)^2,$$

which is equivalent to (13.3).

The first term, t^2, is the component of variance which is explained by the class. The other component, e^2, is unexplained variance. If H_0 is true, the unexplained variance e^2 should be large compared to the explained variance t^2. A standard F-test can be used to compare the two.

It can be shown that t^2 has $g - 1$ degrees of freedom, while e^2 has $M - g$ degrees of freedom, where $M = \sum n_\ell$. Thus the F statistic is given as

$$F = \frac{t^2/(g-1)}{e^2/(M-g)} = \frac{t^2}{e^2} \cdot \frac{M-g}{g-1},$$

which is F distributed with $g - 1$ and $M - g$ degrees of freedom. The p-value is the probability that a random F-distributed variable is greater than or equal to the observed value F as defined above. This probability can be found in a table or by using statistics software. We get the same ordering whether we rank features by F-value or by p-value.

13.3 Feature Subset Selection

Given a set of possible features $\mathcal{F}$, feature selection aims to identify a subset $F \subset \mathcal{F}$ which gives the best possible classification. The features in the optimal subset F do not have to be optimal individually. An extreme and well-known example can be seen in the XOR example from Section 11.2.1. We have two features X_1 and X_2 and a class label $Y = X_1 \oplus X_2$. Let $X_1, X_2 \in \{0, 1\}$ with a 50/50 probability distribution, and both X_1 and X_2 are independent of Y, that is

$$P(X_i = 0) = P(X_i = 0 \mid Y = 0) = P(X_i = 0 \mid Y = 1) = 0.5, \quad \text{for } i = 1, 2.$$

Thus, neither feature X_i is *individually* useful for predicting Y. *Collectively*, in contrast, the joint variable (X_1, X_2) gives complete information about Y, giving a perfect classifier.

In practice, we do not expect such an extreme situation, but it is indeed very common that the usefulness of some features depends heavily on other features included. Two types of dependence are possible. In the XOR example, we see that features reinforce each other. The usefulness of a group of features is more than the sum of its components. Another common situation is where features are redundant, in the sense that the features are highly correlated and contain the same information about the class label. Then the usefulness of the group will be less than the sum of its components.

It is very hard to quantify the usefulness of features and feature vectors, and thus examples necessarily become vague. However, we can note that Cartesian calibration as discussed in Chapter 9 aims to construct features which reinforce each other. The calibrated features are often designed to have the same value for a steganogram and the corresponding cover image, and will then have no discriminatory power at all. Such features are invented only to reinforce the corresponding uncalibrated features. We also have reason to expect many of the features in the well-known feature vectors to be largely redundant. For instance, when we consider large feature vectors where all the features are calculated in a similar way, like SPAM or Shi *et al.*'s Markov features, it is reasonable to expect the individual features to be highly correlated. However, it is not easy to identify which or how many can safely be removed.

Subset selection consists of two sub-problems. Firstly, we need a method to evaluate the usefulness of a given candidate subset, and we will refer to this as subset evaluation. Secondly, we need a method for subset search to traverse possible subsets looking for the best one we can find. We will consider the two sub-problems in turn.

13.3.1 Subset Evaluation

The most obvious way to evaluate a subset of features is to train and test a classifier and use a performance measure such as the accuracy, as the heuristic of the feature subset. This gives rise to so-called *wrapper search*; the feature selection algorithm wraps around the classifier. This approach has the advantage of ranking feature vectors based exactly on how they would perform with the chosen classifier. The down-side is that the training/testing cycle can be quite expensive.

The alternative is so-called *filter search*, where we use some evaluation heuristic independent of the classifier. Normally, such heuristics will be faster than training and testing a classifier, but it may also be an advantage to have a ranking criterion which is independent of the classifier to be used.

The main drawback of wrapper search is that it is slow to train. This is definitely the case for state-of-the-art classifiers like SVM, but there are many classifiers which are faster to train at the expense of accuracy. Such fast classifiers can be used to design a filter search. Thus we would use the evaluation criterion of a wrapper search, but since the intention is to use the selected features with arbitrary classifiers, it is properly described as a filter search. Miche *et al.* (2006) propose a methodology for feature selection in steganalysis, where they use a k nearest neighbours (K-NN) classifier for the filter criterion and SVM for the classification.

The k nearest neighbour algorithm takes no time at all to train; all the calculations are deferred until classification. The entire training set is stored, with a set of feature vectors x_i and corresponding labels y_i. To predict the class of a new feature vector z, the Euclidean distance $\|x_i - z\|$ is calculated for each i, and the k nearest neighbours $i_1, \ldots, i_k$ are identified. The predicted label is taken as the most frequent class label among the k nearest neighbours $y_{i_1}, \ldots, y_{i_k}$.

Clearly, the classification time with k nearest neighbour depends heavily on the size of the training set. The distance calculation is $O(N \cdot M)$, where N is the number of features and M is the number of training vectors. SVM, in contrast, may be slow to train, but the classification time is independent of N. Thus k nearest neighbour is only a good option if the training set can be kept small. We will present a couple of alternative filter criteria later in this chapter, and interested readers can find many more in the literature.

13.3.2 Search Algorithms

Subset search is a general algorithmic problem which appears in many different contexts. Seeking a subset $F \subset \mathcal{F}$ maximising some heuristic $h(F)$, there are $2^{\#\mathcal{F}}$ potential solutions. It is clear that we cannot search them all except in very small cases. Many different algorithms have been designed to search only a subset of the possible solutions in such a way that a near-optimal solution can be expected.

In sequential forward selection, we build our feature set F, starting with an empty set and adding one feature per round. Considering round k, let $X_1, \ldots, X_{k-1}$ be the features already selected. For each candidate feature X, we calculate our heuristic for $\{X, X_1, X_2, \ldots, X_{k-1}\}$. The kth feature X_k is chosen as the one maximising the heuristic.

Where forward selection starts with an empty set which is grown to the desired size, backward selection starts with a large set of features which is pruned by removing one feature every round. Let F_i be the set of all features under consideration in round i. To form the feature set for round $i + 1$, we find X by solving

$$\max_{X \in F_i} h(F_i \setminus \{X\})$$

and set $F_{i+1} = F_i \setminus \{X\}$.

Sequential methods suffer from the fact that once a given feature has been accepted (in forward search) or rejected (in backward search), this decision is final. There is no way to correct a bad choice later. A straight forward solution to this problem is to alternate between forward and backward search, for instance the classic 'plus l, take away r' search, where we alternately make l steps forward and then r steps backward. If we start with an empty set, we need $l > r$. The backward steps can cancel bad choices in previous forward steps and vice versa.

The simple 'plus l, take away r' algorithm can be made more flexible by making l and r dynamic and data dependent. In the classic floating search of Pudil *et al.* (1994), a backward step should only be performed if it leads to an improvement. It works as follows:

1. Given a current set F_k of k features, add the most significant feature $X \notin F_k$ and set $k = k + 1$.

2. Conditionally remove the least significant feature $X \in F_k$, to form F'_{k-1}.
3. If the current subset F'_{k-1} is the best one of size $k - 1$ seen so far, we let $k = k - 1$ and continue removing features by returning to Step 2.
4. Otherwise, we return the conditionally removed feature to the set, and continue adding features by returning to Step 1. (In this case, k and F_k are unchanged since the previous completion of Step 1.)

The search continues until the heuristic converges.

13.4 Selection Using Information Theory

One of the 20th-century milestones in engineering was Claude Shannon's invention of a quantitative measure of *information* (Shannon, 1948). This led to an entirely new field of research, namely information theory. The original application was in communications, but it has come to be used in a range of other areas as well, including machine learning.

13.4.1 Entropy

Consider two stochastic variables X and Y. Information theory asks, how much information does X contain about Y? In communications, Y may be a transmitted message and X is a received message, distorted by noise. In machine learning, it is interesting to ask how much information a feature or feature vector X contains about the class label Y.

Let's first define the *uncertainty* or entropy $H(X)$ of a discrete stochastic variable X, as follows:

$$H(X) = -\sum_{x \in \mathcal{X}} P(X = x) \log P(X = x).$$

Note that the base of the logarithm matters little. Changing the base would only scale the expression by a constant factor. Using base 2, we measure the entropy in bits. With the natural logarithm, we measure it in nats.

Shannon's definition is clearly inspired by the concept of entropy from thermodynamics, but the justification is based on a number of intuitive axioms about uncertainty, or 'lack of information'. Shannon (1948) motivated the definition with three key properties or axioms:

- The entropy is continuous in $P(X = x)$ for each x.
- If the distribution is uniform, then $H(X)$ is a monotonically increasing function in $\#\mathcal{X}$.
- 'If a choice be broken down into two successive choices, the original H should be the weighted sum of the individual values of H.'

The last of these properties may be a little harder to understand than the first two. In order to 'break down' the entropy into successive choices, we need to extend the concept of entropy for joint and conditional distributions. Joint entropy is straightforward. Given two random variables $X_1 \in \mathcal{X}_1$ and $X_2 \in \mathcal{X}_2$, we can consider the joint random variable $(X_1, X_2) \in \mathcal{X}_1 \times \mathcal{X}_2$. Using the joint probability distribution, the joint entropy is given as

$$H(X_1, X_2) = H((X_1, X_2))$$

$$= - \sum_{x,y \in \mathcal{X}_1 \times \mathcal{X}_2} P((X_1, X_2) = (x,y)) \log P((X_1, X_2) = (x,y)).$$

If the two variables X_1 and X_2 are independent, it is trivial to break down the (X_1, X_2) into successive choices:

$$H(X_1, X_2) = - \sum_{x \in \mathcal{X}_1} \sum_{y \in \mathcal{X}_2} P(X_1) P(X_2) \log \big(P(X_1) \cdot P(X_2) \big)$$

$$= - \sum_{x \in \mathcal{X}_1} \sum_{y \in \mathcal{X}_2} P(X_1) P(X_2) \big(\log P(X_1) + \log P(X_2) \big) \qquad (13.4)$$

$$= - \sum_{x \in \mathcal{X}_1} P(X_1) \log P(X_1) - \sum_{y \in \mathcal{X}_2} P(X_2) \log P(X_2)$$

$$= H(X_1) + H(X_2),$$

giving us a sum of individual entropies as required by the axiom.

In the example above, the remaining entropy after observing X_1 is given as $H(X_2)$, but this is only because X_1 and X_2 are independent. Regardless of what we find from observing X_1, the entropy of X_2 is unchanged. In general, however, the entropy of X_2 will change, and vary depending on what X_1 is. This is determined by the conditional probability distribution $P(X_2 \mid X_1 = x_1)$. Every possible value of x_1 gives a different distribution for X_2, each with an entropy given as

$$H(X_2 \mid X_1 = x_1) = - \sum_{x \in \mathcal{X}} P(X_2 = x_2 \mid X_1 = x_1) \log P(X_2 = x_2 \mid X_1 = x_1),$$

where we can see the weighted sum of constituent entropies for different observations of the first variable X_1. We define the conditional entropy $H(X_2 \mid X_1)$ as the expected value of $H(X_2 \mid X_1 = x_1)$ when X_1 is drawn at random from the corresponding probability distribution. In other words,

$$H(X_2 \mid X_1) = - \sum_{x_1 \in \mathcal{X}_1} P(X_1 = x_1) \cdot H(X_2 \mid X_1 = x_1)$$

$$= - \sum_{x_1 \in \mathcal{X}_1} P(x_1) \cdot \sum_{x \in \mathcal{X}_2} P(x_2 \mid x_1) \log P(x_2 \mid x_1).$$

It is straightforward, following the lines of (13.4), to see that

$$H(X_1, X_2) = H(X_2) + H(X_1 \mid X_2) = H(X_1) + H(X_2 \mid X_1),$$

as a Bayes law for entropy. This shows how a choice in general can be broken down into successive choices. The total entropy is given as the unconditional entropy of the first choice and the conditional entropy of the second choice. Moreover, the conditional entropy is the weighted sum (average) of the possible entropies corresponding to different outcomes of the first choice, as Shannon stipulated.

We can also see that a deterministic variable, that is one with a single possible outcome x with $P(X = x) = 1$, has zero entropy as there is no uncertainty in this case. Given a fixed size $\#\mathcal{X}$ we see that $H(X)$ is maximised for the uniform distribution. This is logical because then everything is equally likely, and any attempt at prediction would be pure guesswork, or in other words, maximum uncertainty.

13.4.2 Mutual Information

When the entropy of one variable X_1 changes, from our viewpoint, as a result of observing a different variable X_2, it is reasonable to say that X_2 gives us *information* about X_1. Entropy gives us the means to quantify this information.

The information X_2 gives about X_1 can be defined as the change in the entropy of X_1 as a result of observing X_2, or formally

$$I(X_1; X_2) = H(X_1) - H(X_1|X_2). \tag{13.5}$$

The quantity $I(X_1; X_2)$ is called the *mutual information* between X_1 and X_2, and it is indeed mutual, or symmetric, because

$$I(X_1; X_2) = H(X_1) - H(X_1|X_2) = H(X_1) + H(X_2) - H(X_1, X_2)$$

$$= H(X_2) - H(X_2|X_1) = I(X_2; X_1).$$

This equation is commonly illustrated with the Venn diagram in Figure 13.2. We can also rewrite the definition of $I(X_1; X_2)$ purely in terms of probabilities, as follows:

$$I(X; Y) = \sum_{(x,y)\in\mathcal{X}\times\mathcal{Y}} P(X = x, Y = y) \log \frac{P(X = x, Y = y)}{P(X = x)P(Y = y)}. \tag{13.6}$$

We will also need the conditional mutual information. The definition of $I(X; Y \mid Z = z)$ follows directly from the above, replacing the conditional probability distributions of X and Y given $Z = z$. The conditional mutual information is given as the weighted average, in the same way as we defined conditional entropy:

$$I(X; Y \mid Z) = \sum_{z\in\mathcal{Z}} P(Z = z)I(X; Y \mid Z = z).$$

The entropy and mutual information are defined from the probability distribution, which is a theoretical and unobservable concept. In practice, we will have to estimate the quantities. In the discrete case, this is straightforward and uncontroversial.

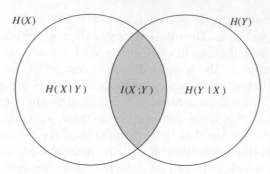

Figure 13.2 Venn diagram showing the various entropies and mutual information of two random variables X and Y

We simply use relative frequencies to estimate probabilities, and calculate the entropy and mutual information substituting estimates for the probabilities.

Example 13.4.1 (The XOR problem) *Recall the XOR example where we have three random variables $X, Y, Z \in \{0, 1\}$, with X and Y independent and $Z = X \oplus Y$. Thus each variable is given as the XOR of the other two. Assume that each variable is uniformly distributed. The three variables have the same entropy,*

$$H(X) = H(Y) = H(Z) = 2 \cdot (-0.5) \cdot \log 0.5 = 1.$$

Each pair (X, Y), (X, Z) and (Y, Z) is uniformly distributed on the set $\{0, 1\}^2$. Thus, the joint entropy is given as

$$H(X, Y) = H(X, Z) = H(Y, Z) = 4 \cdot (-0.25) \cdot \log 0.25 = 2.$$

It follows that the mutual information is given as

$$I(X, Z) = H(X) + H(Z) - H(X, Z) = 1 + 1 - 2 = 0.$$

In other words, X contains zero information about Z, confirming what we argued previously, that X is not individually a useful feature to predict Z.

There is a strong relationship between mutual information and entropy on the one hand, and classifier performance on the other.

Theorem 13.4.2 *The Bayes error of a classifier is bounded as $\mathsf{P}(g(X) \neq Y) \leq \frac{1}{2}H(Y|X)$, where $g(X)$ is the predicted class for a given feature vector X.*

Note that the entropy $H(Y|X)$ is the regular, discrete entropy of Y. It is only the conditional variable X which is continuous, but $H(Y|X)$ is still well-defined as the expected value of $H(Y|X = x)$.

Proof. The Bayes error is

$$P_E = P(h(X) \neq Y) = \sum_y \int_{x \in R_y} P(Y \neq y|x)f(x)dx,$$

where $R_y = \{x|h(x) = y\}$ is the set of x observations that would be classified as class y. The entropy, in contrast is given as

$$H(Y \mid X) = -\sum_y \int P(y|x) \log P(y \mid x)f(x)dx.$$

To prove the theorem, we need to show that

$$0 \leq \sum_y \int f(x)P(y|x)\left[\frac{-1}{2} \log P(y \mid x) - g(x,y)\right]dx,$$

where $g(x,y) = 0$ for $y \in R_y$ and $g(x,y) = 1$ otherwise. It is sufficient to show for all x that

$$0 \leq \sum_y P(y|x)\left[\frac{-1}{2} \log P(y \mid x) - g(x,y)\right].$$

Let $p_x = P(y|x)$ such that $x \in R_y$, i.e. p_x is the posterior probability that the predicted class be correct. Recall that R_y is defined so that P_E is minimised, and therefore

$$\sum_y P(y|x)g(x,y) = 1 - p_x.$$

The entropy is bounded as

$$\sum_y P(y|x)\frac{-1}{2} \log P(y \mid x) \geq \frac{-1}{2} \log p_x,$$

since p_x is the largest conditional probability $P(y|x)$ for any y. It can easily be shown that $(\log p_x)/2 \geq 1 - p_x$ $(0 \leq p_x \leq 1)$, completing the proof.

13.4.3 Multivariate Information

To present the general framework for feature selection using information theory, we will need multivariate mutual information. This can be defined in many different ways. Yeung (2002) develops information theory in terms of measure theory, where entropy, mutual information and multivariate information are special cases of the same measure. As beautiful as this framework is, there is no room for the mathematical detail in this book. We give just the definition of multivariate information, as follows.

Definition 13.4.3 *Multivariate mutual information is defined for any set S of stochastic variables, as follows:*

$$I(S) = -\sum_{T \subset S}(-1)^{\#S \backslash T}H(T).$$

We can clearly see traces of the inclusion/exclusion theorem applied to some measure $H = I$. It is easily checked that mutual information is a special case where $S = \{X, Y\}$ by noting that

$$I(S) = H(X) + H(Y) - H(\{X, Y\}) = I(X; Y).$$

We note that the definition also applies to singleton and empty sets, where $I(\{X\}) = H(X)$ and $I(\emptyset) = 0$. The first author to generalise mutual information for three or more variables was McGill (1954), who called it interaction information. His definition is quoted below, but it can be shown that interaction information and multivariate mutual information are equivalent (Yeung, 2002). We will use both definitions in the sequel.

Definition 13.4.4 (Interaction information (McGill, 1954)) *The* interaction informa- tion *of a set of random variables* $\{X_1, X_2, \ldots, X_N\}$ *for $N \geq 3$ is defined recursively as*

$$I(\{X_1, X_2, \ldots, X_N\}) = I(\{X_1, X_2, \ldots, X_{N-1}\} \mid X_N) - I(\{X_1, X_2, \ldots, X_{N-1}\}),$$

where the conditional form is found simply by marginalising over the distribution of X_N. For $N = 2$, we define $I(\{X_1, X_2\}) = I(X_1; X_2)$.

Note that $S = \{X, Y\}$ is not in any way considered as a joint stochastic variable (X, Y). Considering a single joint variable (X, Y) gives the interaction information

$$I(\{(X, Y)\}) = H(X, Y),$$

which is the joint entropy, while $S = \{X, Y\}$ leads to

$$I(\{X, Y\}) = H(X) + H(Y) - H(X, Y) = I(X; Y),$$

which is the mutual information. Considering three variables, the interaction infor- mation is given as

$$I(\{X, Y, Z\}) = -H(X) - H(Y) - H(Z) + H(X, Y) + H(X, Z) + H(Y, Z) - H(X, Y, Z),$$

which does not coincide with the mutual information $I(X, Y; Z)$, nor with $I(X; Y, Z)$. The relationship is visualised in Figure 13.3.

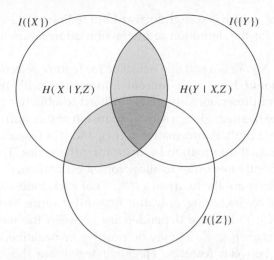

Figure 13.3 Venn diagram showing interaction information for three random variables. The dark grey area is the third-order interaction information $I(\{X, Y, Z\})$. The total grey area (dark and light) is the mutual information $I(X; Y, Z)$

Example 13.4.5 *Consider again the XOR example (Example 13.4.1). The multivariate information is given as*

$$I(\{X, Y, Z\}) = -H(X) - H(Y) - H(Z)$$
$$+ H(X, Y) + H(X, Z) + H(Y, Z) - H(X, Y, Z)$$
$$= -3 + 6 - 2 = +1.$$

Using McGill's definition, we see that when Z is known, X gives complete information about Y, whereas with Z unknown we get no information. Thus

$$I(\{X, Y, Z\}) = I(X; Y \mid Z) - I(X; Y) = 1 - 0 = 1,$$

which is the same result, as expected.

Our interest in the multivariate mutual information comes from the fact that we can use it to estimate the pairwise mutual information of joint stochastic variables. The following theorem is taken from Brown (2009).

Theorem 13.4.6 *Given a set of features $S = \{X_1, \ldots, X_N\}$ and a class label Y, their mutual information can be expanded as*

$$I(X_1, \ldots, X_N; Y) = \sum_{T \subset S, T \neq \emptyset} I(T \cup Y).$$

We omit the proof, which is straightforward but tedious. An interested reader can find it by inserting for the definition of interaction information in each term on the right-hand side.

Clearly, $I(X_1, \ldots, X_N; Y)$ is a suitable heuristic for feature selection. A good feature set $(X_1, \ldots, X_N)$ should have high mutual information with the class label. For large N, the curse of dimensionality makes it hard to obtain a robust estimate for $I(X_1, \ldots, X_N; Y)$. The theorem gives a series expansion of the mutual information into a sum of multivariate mutual information terms, and it is reasonable and natural to approximate the mutual information by truncating the series. Thus we may be left with terms of sufficiently low order to allow robust estimation.

The first-order terms are the heuristics $I(X_i; Y)$ of each individual feature X_i, and they are non-negative. Including only the first-order terms would tend to over-estimate $I(X_1, \ldots, X_N; Y)$ because dependencies between the features are ignored. The second-order terms $I(X_i, X_j; Y)$ may be positive or negative, and measure pair-wise relationships between features. Negative terms are the result of pair-wise redundancy between features. A positive second-order term, as for instance in the XOR example, indicates that two features reinforce each other. Higher-order terms measure relationships within larger groups of features.

13.4.4 Information Theory with Continuous Sets

In most cases, our features are real or floating point numbers, and the discrete entropy we have discussed so far is not immediately applicable. One common method is to divide the range of the features into a finite collection of bins. Identifying each feature value with its bin, we get a discrete distribution, and discrete entropy and mutual information can be used. We call this *binning*. Alternatively, it is possible to extend the definition of entropy and information for continuous random variables, using integrals in lieu of sums. This turns out very well for mutual information, but not as well for entropy.

The *differential entropy* is defined as a natural continuous set analogue of the discrete entropy:

$$H(X) = -\int_{\mathcal{X}} f(x) \log f(x) dx, \tag{13.7}$$

where $f(x)$ is the probability density function. Although this definition is both natural and intuitive, it does not have the same nice properties as the discrete entropy. Probabilities are bounded between zero and one and guarantee that the discrete entropy is non-negative. Probability densities, in contrast, are unbounded above, and this may give rise to a negative differential entropy.

Another problem is that the differential entropy is sensitive to scaling of the sample space. Replacing x by $y = \alpha x$ for $\alpha > 1$ will make the distribution flatter, so the density $f(y)$ will be lower and the entropy will be larger. The same problem is seen with binned variables, where the entropy will increase with the number of bins. Intuitively, one would think that the uncertainty and the information should be independent of the unit of measurement, so this sensitivity to scaling may make differential entropy unsuitable.

Unlike differential entropy, continuous mutual information can be shown to retain the interpretations of discrete mutual information. It can be derived either from $I(X; Y) = H(X) - H(X|Y)$ using the definition of differential entropy (13.7), or equivalently as a natural analogue of the discrete formula (13.6). Thus we can write

$$I(X, Y) = \int_y \int_x f(x, y) \log \frac{f(x, y)}{f_X(x) f_Y(y)} dx dy. \tag{13.8}$$

It is possible to show that this definition is the limit of discrete mutual information with binned variables as the bin size decreases. One can see this intuitively, by noting that the scaling factors in the numerator and denominator cancel out. Thus the continuous mutual information $I(X; Y)$ between a feature X and the class label Y is a potentially good heuristic for the classification ability, even if continuous entropy is not well-defined.

13.4.5 Estimation of Entropy and Information

The problem of estimating entropy and information is closely related to that of estimating probability density functions (Section 12.2). The simplest approach is to use binning to discretise the random variable as addressed above. This corresponds to using the histogram to estimate the PDF. This approach is dominant in the literature on machine learning and it is rarely subject to discussion. The challenge is, both for density estimation and for mutual information, to choose the right bin width. We can use the same advice on bin width selection for entropy estimation as for density estimation. Sturges' rule using $1 + \log_2 N$ bins is a good starting point, but in the end it boils down to trial and error to find the ideal bin width.

For the purpose of feature selection, our main interest is in the mutual and interaction information, and not in the entropy per se. We can estimate information using any of the rules

$$\hat{I}(X; Y) = \hat{H}(X) + \hat{H}(Y) - \hat{H}(X, Y), \tag{13.9}$$

$$\hat{I}(X; Y) = \hat{H}(X) - \hat{H}(X \mid Y), \tag{13.10}$$

$$\hat{I}(X, Y) = \sum_{(x,y) \in \mathcal{X} \times \mathcal{Y}} \hat{P}(X = x, X = y) \log \frac{\hat{P}(X = x, X = y)}{\hat{P}(X = x) \hat{P}(Y = y)}, \tag{13.11}$$

where $\hat{P}$ and $\hat{H}$ are estimates of the probability density and the entropy respectively. In principle, they all give the same results but they may very well have different error margins.

Bin size selection is a particularly obvious problem in entropy estimation, as the entropy will increase as a function of the number of bins. Thus, entropy estimates calculated with different bin sizes may not be comparable. Estimating mutual information, one should use the same bin boundaries on each axis for all the estimates used, although X and Y may be binned differently. We will investigate this further in the context of continuous information.

There are several methods to estimate differential entropy and continuous mutual information. The one which is intuitively easiest to understand is the entropy estimate of Ahmad and Lin (1976). Then either (13.9) or (13.10) can be used, depending on whether the joint or conditional entropy is easiest to estimate. For the purpose of this book, we will be content with this simple approach, which relies on kernel density estimators as discovered in Section 12.2.2. Interested readers may consult Kraskov *et al.* (2004) for a more recent take on mutual information estimation, and an estimator based on kth nearest neighbour estimation.

The Ahmad–Lin estimator is given as

$$\hat{H}(X) = \frac{1}{\#\mathcal{S}} \sum_{x \in \mathcal{S}} \log \hat{f}(x),$$

where $\mathcal{S}$ is the set of observations of X and $\hat{f}$ is the kernel density estimate. An implementation is shown in Code Example 13.1. The term $f(x)$ in the definition of differential entropy does not occur explicitly in the estimator. However, we sample $\log \hat{f}(x)$ at random points x drawn according to $f(x)$, and thus the $f(x)$ term is represented implicitly as the sum.

13.4.6 Ranking Features

We have already argued that $I(X_1, \ldots, X_N; Y)$ is a useful evaluation heuristic for a set of features $X_1, \ldots, X_N$ and a class label Y. It is a very intuitive measure, as it quantifies the information about Y contained in the features. The larger the mutual information, the better classification accuracy can be expected. Now we turn to the computational problem of estimating this information.

For large N, the curse of dimensionality makes it impractical to calculate $I(X_1, \ldots, X_N)$. We can approximate it by a truncation of the series expansion in

---

Code Example 13.1 The Ahmad–Lin estimator in Python, using the Gaussian KDE from scipy

---

```
import numpy as np
from scipy.stats import gaussian_kde as kde
def ahmadlin(X):
    f = kde(X)
    y = np.log( f.evaluate(X) )
    return = - sum(y)/len(y)
```

Theorem 13.4.6. It is common to use the second-order approximation

$$I(X_1,\ldots,X_N;Y) \approx \sum_{i=1}^{N} I(X_i;Y) + \sum_{\substack{i,j=1 \\ i<j}}^{N} I(\{X_i,X_j,Y\}). \tag{13.12}$$

At worst, only three-dimensional sample spaces have to be calculated. This can be used as a filter criterion as it stands, but it is still computationally expensive for large N. In sequential forward selection, it is possible to simplify the expression by reusing computations from previous rounds.

In the N-th round we compare subsets where $X_1,\ldots,X_{N-1}$ have been fixed by the previous round, and only X_N is to be selected. If we write I_{N-1}^* for the 'winning' heuristic from round $N-1$, we can rewrite (13.12) as

$$I(X_1,\ldots,X_N;Y) \approx I_{N-1}^* + I(X_N;Y) + \sum_{i=1}^{N-1} I(\{X_i,X_N,Y\}). \tag{13.13}$$

Since I_{N-1}^* is constant, the interesting heuristic to evaluate a single candidate feature can be written as

$$\begin{aligned} J_{\text{FOU}}(X_N) &= I(X_N;Y) + \sum_{i=1}^{N-1} I(\{X_i,X_N,Y\}) \\ &= I(X_N;Y) - \sum_{i=1}^{N-1} I(X_i;X_N) + \sum_{i=1}^{N-1} I(X_i;X_N \mid Y), \end{aligned} \tag{13.14}$$

where the last equality follows from McGill's definition (Definition 13.4.4). Brown (2009) calls this the *first-order utility* (FOU). Note that only $I(X_{N-1};X_N)$ and $I(X_{N-1};X_N \mid Y)$ need to be estimated in round N. All the other terms have already been calculated in previous rounds.

In order to estimate J_{FOU}, we need to estimate both unconditional and conditional mutual information. Recall that Y is discrete and usually drawn from a small set of class labels, thus we can write

$$\hat{I}(X_i;X_N \mid Y) = \sum_{y \in \mathcal{Y}} P(Y=y)\hat{I}(X_i;X_N \mid Y=y),$$

where $\hat{I}(X_i;X_N \mid Y=y)$ can be calculated as the estimated mutual information considering the samples from class y only.

Many of the feature selection heuristics in the literature are variations of the FOU formula. A great deal of them can be written in the form

$$J = I(X;Y) - \beta \sum_{k=1}^{n-1} I(X_n;X_k) + \gamma \sum_{k=1}^{n-1} I(X_n;X_k|Y), \tag{13.15}$$

by varying the parameters (β, γ). For $\beta = \gamma = 1$ we obviously get the FOU. Brown (2009) recommends using joint mutual information (JMI), which we define as

$$J_{\mathrm{JMI}} = I(X_n; Y) - \frac{1}{n-1} \sum_{k=1}^{n-1} (I(X_n; X_k) - I(X_n; X_k \mid Y)).$$

In other words, we use $\alpha = \gamma = 1/(n-1)$ in (13.15). For $n = 1$, we let $J_{\mathrm{JMI}} = I(X_n; Y)$.

The JMI criterion was originally introduced by Yang and Moody (1999), who defined the JMI heuristic as

$$J'_{\mathrm{JMI}} = \sum_{k=1}^{n-1} I(X_n, X_k; Y). \tag{13.16}$$

Note that even though $J'_{\mathrm{JMI}} \neq J_{\mathrm{JMI}}$, they both give the same ranking of features for $N \geq 2$. In fact,

$$J'_{\mathrm{JMI}} = (n-1)\left(J_{\mathrm{JMI}} + \sum_{i=1}^{N-1} I(X_i; Y) \right).$$

Both the last term and the multiplication factor are constant for all features X_N considered in round N. For $n = 1$, $J'_{\mathrm{JMI}} \equiv 0$ cannot rank the features, and one would use $I(X_1; Y) = J_{\mathrm{JMI}}$ as the criterion in the first round.

In order to calculate J'_{JMI} empirically, we note that the first $n - 2$ terms $I(X_i, X_n; Y)$ in the sum are the score from the previous round. We only need to calculate

$$\hat{I}(X_{n-1}, X_n; Y) = \hat{H}(X_{n-1}, X_n) - \hat{H}(X_{n-1}, X_n \mid Y).$$

For a two-class problem with no class skew, this gives

$$\hat{I}(X_{n-1}, X_n; Y) = \hat{H}(X_{n-1}, X_n) - 0.5\hat{H}(X_{n-1}, X_n \mid Y = 0)$$
$$- 0.5\hat{H}(X_{n-1}, X_n \mid Y = 1).$$

Some care must be taken to ensure that the entropy estimates are compatible. A working approach based on the Ahmad–Lin estimator is to scale all the features to have unit variance and make sure to use the same bandwidth matrix for the KDE for all three entropy estimates.

An implementation of JMI selection is shown in Code Example 13.2. An object is instantiated with a 2-D array of feature vectors T and a 1-D array of class labels Y. The object has a state maintaining a list of features selected thus far, and a list of evaluation scores from the previous round. The _getRankData() method evaluates a single feature, and next() returns the best feature for the current round and proceeds to the next round. Note that we use the KDE class from Code Example 12.2, so that we can explicitly use the same bandwidth matrix for all three KDEs. The entropy h2 and conditional entropy c2 are calculated with Ahmad and Lin's formula as before. The computational cost is dominated by the evaluation of three KDEs per round.

Code Example 13.2 A class to do JMI feature selection using Yang and Moody's formula. Note that KDE is the class shown in Code Example 12.2. The class interface is explained in the text

```python
import numpy as np
class JMI(object):
    def _ _init_ _(self,Y,T ):
        (self.dim,self.size) = T.shape
        self.score = None
        self.selected = []
        (self.Y, self.T) = (Y,T)
    def next(self,*a,**kw):
        R = map( self._getRankData, xrange( self.dim ) )
        if len(self.selected) > 0: self.score = [ y for (x,y) in R ]
        R.sort( cmp=lambda x,y : cmp(x[1],y[1]) )
        ( gamma, score ) = R[0]      # Pick the winner
        self.selected.append( gamma )
        return gamma
    def _getRankData( self, gamma ):
        idx = [gamma]
        if len(self.selected) > 0: idx.append( self.selected[−1] )
        # Get score from previous round:
        if self.score != None: R = self.score[gamma]
        else: R = 0
        # Calculate kernel density estimates
        C = self.T[:,(self.Y == 0)]   # Select clean images
        S = self.T[:,(self.Y == 1)]   # Select steganograms
        K = KDE( self.T[idx,:] )
        KC = KDE( C[idx,:], bandwidth=K.bandwidth )
        KS = KDE( S[idx,:], bandwidth=K.bandwidth )
        # Calculate empirical entropies
        (nC,d) = C.shape
        (nS,d) = S.shape
        h2 = − sum( np.log( K.evaluate(X) ) ) / (nC+nS)
        HC = − sum( np.log( KC.evaluate(X) ) ) / nC
        HS = − sum( np.log( KS.evaluate(X) ) ) / nS
        c2 = ( n1*HC + n2*HS ) / (n1+n2)
        # Return the feature index and its score
        return ( gamma, R + h2 − c2 )
```

13.5 Boosting Feature Selection

Boosting feature selection (BFS) was introduced by Tieu and Viola (2004), and Dong *et al.* (2008) proposed using it in the context of steganalysis. BFS builds on a previous idea of *boosting*, which is used to design a classifier as a linear combination of multiple constituent classifiers. To use boosting for feature selection, we consider each individual feature x_i as a classification heuristic, where negative and positive values of x_i are taken to indicate different classes. The BFS algorithm combines this idea with forward sequential search.

The BFS classifier after $m - 1$ rounds of forward search is written as

$$F_{m-1}(\mathbf{x}) = \sum_{i=1}^{n} \beta_i x_i,$$

for any point $\mathbf{x}$ in feature space; where $\boldsymbol{\beta} = (\beta_1, \ldots, \beta_n)$ has at most $m - 1$ elements not equal to 0. Note that before the first round, the above formula gives the 'empty' classifier $F_0 \equiv 0$. In round m we seek to add one new feature to the classifier, so that the new classifier will be given as

$$F_m(\mathbf{x}) = F_{m-1}(\mathbf{x}) + \beta x_\gamma,$$

where γ and β are to be optimised.

Let $\mathbf{x}_i$ for $i = 1, \ldots, N$ be the training feature vectors, and let $y_i = \pm 1$ be the associated class labels. In each round, we start with a classifier F_{m-1} and define a new classifier F_m by selecting a new feature x_γ and updating the associated weight β_γ. Each feature is selected by minimising the squared error of the classifier F_m, i.e.

$$\min_{\beta, \gamma} E_m(\gamma, \beta) = \sum_i \cdot [y_i - F_{m-1}(\mathbf{x}_i) - \beta x_{i,\gamma}]^2, \qquad (13.17)$$

where $x_{i,\gamma}$ is the γth feature of the ith training vector. Let (β, γ) be the solution of the minimisation problem. The index γ identifies a feature to be included in the selection, whereas β represents an update to the classifier. The classifier F_m for the next round is defined by reassigning the weight vector $\boldsymbol{\beta}$ by setting $\beta_\gamma := \beta_\gamma + \beta$. The process can be repeated until the square error converges.

The cost function E_m to be optimised in (13.17) is obviously not continuous in the discrete variable γ, so the easiest way to optimise it is to solve

$$\min_{\beta} E_m(\beta, \gamma) = \sum_i \cdot [y_i - F_{m-1}(\mathbf{x}_i) - \beta x_{i,\gamma}]^2,$$

for each $\gamma = 1, 2, \ldots, N$. This gives a heuristic $E_m(\beta(\gamma), \gamma)$ ranking the features, and we use the feature γ with the lowest heuristic value as the mth feature for our classifier. The corresponding weighting $\beta(\gamma)$ is the weighting of x_γ in the updated BFS classifier.

Code Example 13.3 A class to do BFS feature selection. The interface mimics that from Code Example 13.2

```
import numpy as np
class BFS(object):
    def _ _init_ _(self,Y,T ):
        (self.dim,self.size) = T.shape
        self.beta = np.zeros( self.dim )
        (self.Y, self.T) = (Y,T)
    def next(self):
        R = map( self._getRankData, xrange( self.dim ) )
        R.sort( cmp=lambda x,y : cmp(x[1],y[1]) )
        selected = R[0]
        self.beta[selected[0]] += selected[2]
        return selected[0]
    def _getRankData( self, gamma ):
        F = np.dot( self.beta, self.T )      # Prediction
        W = self.Y - F                       # Error
        b = self.T[gamma,:]                  # current feature
        beta = sum( W * b ) / sum( b**2 )    # New beta value
        T = W - beta * self.T[gamma,:]       # New error
        E = sum( T**2 ) / len(self.Y)        # Total squared error
        return ( gamma, E, beta )
```

Code Example 13.3 shows how this can be done in Python using roughly the same interface as in the previous JMI implementation. A BFS object is instantiated by providing the class labels as a 1-D numpy array, and the feature vectors as the rows of an $N \times l$ numpy array. Initially the BFS object is in round $m = 0$. The next() method returns the next feature to be selected, and the state proceeds to round $m + 1$, including updating the internal classifier F_m.

Remark 13.1 *Note that BFS can be used both as a filter method for feature selection for an arbitrary classifier algorithm, and as a classifier in its own right. The function F_m, or rather sign $F_m(x)$ to be precise, constitutes the trained classifier. To use it just for feature selection, we select the features x_i that have non-zero weighting $\beta_i \neq 0$.*

13.6 Applications in Steganalysis

There are a few, but only a few, examples of feature selection being used in steganalysis. Pevný *et al.* (2009a) even state that there is no known case in steganalysis where feature selection improves the classifier. In a massive effort to break the HUGO system, Fridrich *et al.* (2011) ended up with a massive 24 993-dimensional feature vector, and they argue that high dimension poses no harm.

Part of the problem is that the size of the training set must correspond to the feature dimensionality. In the attack on HUGO, Fridrich *et al.* used a training set of 30000–90000 images. This poses two challenges. Firstly, the task of collecting a diverse and representative image base of tens of thousands of images is difficult and time-consuming, especially if multiple classes of images are going to be considered and compared. Secondly, the computational cost of training the classifier increases with the size of the training set. In fact, Fridrich *et al.* gave up SVM in favour of faster algorithms when they took up using tens of thousands of features.

Most authors use both shorter feature vectors and smaller training sets. If we try to combine many features from different authors, without having the opportunity to increase the training set correspondingly, then we will quite likely have a case for feature selection. *A priori*, there is no reason to expect the published feature vectors to be optimal selections. It is possible that better classifiers are available by combining existing features into new feature vectors. This idea is known as feature-level fusion. An interesting exercise is to fuse all known feature vectors into one, and then use feature selection techniques to find an optimal one. A pioneer in using feature-level fusion in steganalysis is Dong *et al.* (2008), who combined it with BFS.

The possible use of feature selection for the purpose of interpretability has been explored by Miche *et al.* (2009). They described the objective as a reverse engineering exercise, where the most useful features are used to identify artifacts caused by embedding and thus help in reverse engineering the stego-system. A number of stego-systems were tested, identifying the features which are sensitive to each system. It seems that their work only scratched the surface, and they did not cover all of the current stego-systems. It should be interesting to see more research along this line.

13.6.1 Correlation Coefficient

Pevný *et al.* (2009a) included experiments with feature selection from the SPAM features. Their approach goes a bit beyond the framework that we have discussed so far, in that their heuristic is based not on the class label, but on the number of coefficients changed by the embedding. In a sense, this tunes the feature selection for quantitative steganalysis using regression, rather than to binary classification. Still, features which are useful for one problem are likely to be useful for the other.

Let Q denote the number of embedding changes in the intercepted image. For each feature X_i, the *correlation coefficient* is defined as

$$\mathrm{corr}(X_i, Q) = \frac{\mathrm{Cov}(X_i, Q)}{\sqrt{\mathrm{Var}(X_i)\,\mathrm{Var}(Q)}}$$

$$= \frac{\mathrm{E}(X_i Q) - \mathrm{E}(X_i)\mathrm{E}(Q)}{\sqrt{\mathrm{E}(X_i^2) - \mathrm{E}(X_i)^2} \cdot \sqrt{\mathrm{E}(Q^2) - \mathrm{E}(Q)^2}}.$$

The sample correlation coefficient can be calculated using the sample means in lieu of the expectations, and this is used to rank the features and select those with the largest correlation coefficient. Evidently, one can replace Q by a numeric class label, but we have not seen this approach tested.

Pevný *et al.* (2009a) reports that the accuracy increases until 200 features, and then it stabilises. In other words, they find no evidence of overfitting where feature selection can increase the accuracy in the SPAM features. Furthermore, the optimal features depend on the data set used. Across four image databases considered, 114 out of the 200 best features are shared for all image sets. Thus one might conclude that the full SPAM feature vector should be used for a universal steganalyser.

13.6.2 Optimised Feature Vectors for JPEG

In an attempt to make the optimal feature vector for JPEG, based on all the different features we have discussed, we have run a number of experiments using feature selection. We used the same training and test sets as we have used throughout, with 1000 training images and 4000 test images from the BOWS collection. To do the feature selection, we used the training set with 500 clean images JPEG compressed at QF75 and 500 steganograms with 512 byte messages embedded with F5. A total of about 6700 features had been implemented and used. Our experiments closely follow those reported by Schaathun (2011), except for a few minor modifications and bug fixes.

We have selected features using FOU, JMI and BFS. Figure 13.4 shows the relationship between dimensionality and testing accuracy, where the best features have been used according to each of the selection criteria. It is interesting to see how quickly the curve flattens, showing that decent accuracy can be achieved with less than 50 features, and little is gained by adding features beyond 100. Recalling that

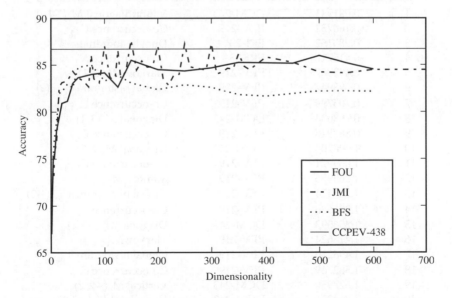

Figure 13.4   Accuracy for SVM using automatically selected features

NCPEV-219 and CCPEV-438 had accuracies of 86.3% and 86.7% respectively, we note that automated feature selection does not improve the accuracy compared to state-of-the-art feature vectors.

We note that there is no evidence of the curse of dimensionality kicking in within the 700 dimensions we have covered. In fact, both the training and testing error rates appear to stabilise. We also ran a comparison test using all the JPEG-based features (CCPEV-438, FRI-23, Markov-243, HCF-390 and CCCP-54) to see how much trouble the dimensionality can cause. It gave an accuracy of 87.2%, which is better than CCPEV-438, but not significantly so.

Even though automated feature selection does not improve the accuracy, improvements may be found by studying which features are selected. In Table 13.1 we have listed the first 20 features in the order they were selected with the JMI measure. The DCM-243 feature vector is Shi *et al.*'s Markov-based features subject to difference calibration. As we see, different variations of Markov-based features dominate the list, with a good number of co-occurrence and global histogram features. The single dual histogram feature is the odd one out. Continuing down the list, Markov-based features continue to dominate with local histogram features as a clear number two.

There is no obvious pattern to which individual features are selected from each feature group. Since all of these features are representative of fairly large groups, it is tempting to see if these groups form good feature vectors in their own right. The results are shown in Table 13.2. We note that just the Markov-based features from

Table 13.1 Top 20 features in the selection according to JMI

No.	JMI score	Vector	Feature
1	0.081941	DCM-243	Minor diagonal $M_v(-1,0)$
2	0.160732	PEV-219	Co-occurrence $C_{-1,-1}$
3	0.283152	FRI-23	Dual histogram $\mathbf{g}_0$
4	0.415717	FRI-23	Co-occurrence N_2
5	0.492367	PEV-219	Markov $M(-2,1)$
6	0.598947	PEV-219	Global histogram $h_{\text{glob}}(2)$
7	0.705799	PEV-219	Co-occurrence $C_{-1,1}$
8	0.793034	DCM-243	Diagonal $M_d(3,1)$
9	0.887945	PEV-219	Co-occurrence $C_{1,1}$
10	1.000203	DCM-243	Diagonal $M_d(2,1)$
11	1.069971	PEV-219	Co-occurrence $C_{1,-1}$
12	1.205956	PEV-219	Markov $M_{2,1}$
13	1.265424	PEV-219	Global histogram $h_{\text{glob}}(-2)$
14	1.296686	PEV-219	Co-occurrence $C_{-2,0}$
15	1.356093	DCM-243	Diagonal $M_d(-2,1)$
16	1.418624	PEV-219	Markov $M_{3,1}$
17	1.498236	PEV-219	Global histogram $h_{\text{glob}}(1)$
18	1.582769	PEV-219	Co-occurrence $C_{2,0}$
19	1.627950	DCM-243	Vertical $M_v(-2,2)$
20	1.764276	DCM-243	Horizontal $M_h(-2,1)$

Table 13.2 Accuracies for some new feature vectors. Elementary subsets from known feature vectors are listed in the first part. New combinations are listed in the lower part

Name	Accuracy	Comprises
CCPM-162	81.9	Markov features from CCPEV-438
NCPM-81	81.0	Markov features from NCPEV-219
CCCM-50	74.5	Co-occurrence matrix from CCPEV-438
DCCM-25	69.5	Co-occurrence matrix from PEV-219
NCCM-25	68.5	Co-occurrence matrix from NCPEV-219
HGS-162	90.3	HGS-160 + variation from CCPEV-438
HGS-178	90.1	HGS-142 + local histogram $h_{i,j}(\pm 5)$ (CCPEV-438)
HGS-160	90.1	HGS-142 + local histogram $h_{i,j}(\pm 5)$ (NCPEV-219)
HGS-180	89.9	HGS-178 + variation from CCPEV-438
HGS-212	89.4	CCPM-162 + CCCM50
HGS-142	88.7	HGS-131 + global histogram from PEV-219
HGS-131	88.2	NCPM-81 + CCCM50

NCPEV-219 or CCPEV-438 give decent classification, and combining the Markov and co-occurrence features from CCPEV-438, HGS-212 gives a statistically significant improvement in the accuracy with less than half the dimensionality.

The accuracy can be improved further by including other features as shown in the table. The local histogram features used in HGS-160 and HGS-178 are the ± 5 bins from each of the nine frequencies used, taken from NCPEV-219 and CCPEV-438, respectively. Although automated feature selection does not always manage to improve on the accuracy, we have demonstrated that feature selection methods can give very useful guidance to manual selection of feature vectors. Schaathun (2011) also reported consistent results when training and testing the same feature selection with Outguess 0.1 at quality factor 75 and with F5 at quality factor 85.

14

The Steganalysis Problem

So far, we have defined steganalysis as the eavesdropper's problem, deciding whether or not a given information object contains covert information in addition to the overt information. In reality, the eavesdropper's problem is not so homogeneous. Different scenarios may require different types of steganalysis, and sometimes the eavesdropper needs more information about the covert message.

Generally, the difficulty of steganalysis decreases with the size of the covert message. An important question from the steganographer's point of view is 'how much data can be embedded and still remain undetectable?' This maximum quantity of undetectable information may be called the steganographic *capacity*. The concept of *undetectable* is a fuzzy concept, giving rise to a number of proposed *security criteria*.

In this chapter, we will discuss various use cases and how they call for different properties in steganalysers. We will also see how composite steganalytic systems can be designed, using multiple constituent steganalysers.

14.1 Different Use Cases

14.1.1 Who are Alice and Bob?

In the original phrasing of the problem, Alice and Bob are prisoners who want to use steganography to coordinate escape plans under the nose of Wendy. Already arrested, Alice and Bob are not necessarily assumed to be innocent. It is quite possible that Wendy does not need very strong evidence to withdraw their privilege by closing the communication line between Alice and Bob, and in any case she can dedicate special resources to the monitoring of their particular communication line. We will continue to refer to this scenario as the *prisoners' problem*.

Other user scenarios may be more or less different from the textbook scenario. Let us first consider a fairly similar one. Alice and Bob are still individuals, transmitting steganograms in a one-to-one conversation, but they are free citizens and not subject

Machine Learning in Image Steganalysis, First Edition. Hans Georg Schaathun.
© 2012 John Wiley & Sons, Ltd. Published 2012 by John Wiley & Sons, Ltd.

to any special legal limitations. They could be involved in organised crime, or they could be intelligence agents of a foreign power. In this case Wendy will need much stronger evidence in order to punish or restrain Alice and Bob; she must prove, beyond any reasonable doubt, that Alice and Bob are breaking the law. Furthermore, Alice and Bob may not receive any special attention from Wendy, as she must split her resources between a number of potential suspects. We will refer to this scenario as *one-to-one steganography*.

A popular, alleged scenario is Alice using steganography in a kind of broadcast dead-drop. Dead-drops traditionally refer to hiding places in public locations, where one agent could hide a message to be picked up by another at a later stage. In steganography, we imagine dead-drops on public Internet sites, where innocent-looking images can be posted, conveying secret messages to anyone who knows the secret key. Clearly, the intended receiver may be an individual, Bob (as before), or a large network of fellow conspirators. If the steganogram is posted on a site like eBay or Facebook, a lot of innocent and unknowing users are likely to see the image. Even when Alice is suspected of a conspiracy, Bob may not become more suspicious than a myriad other users downloading the same image. We call this scenario *dead-drop steganography*.

In all the scenarios mentioned so far, the message may be long or short. Alice and Bob may need to exchange maps, floor plans, and images, requiring several images at full capacity to hide the entire message. It is also possible that the message is very short, like 'go tomorrow 8pm', if the context is already known and shared by Alice and Bob. As we have seen, very short messages are almost impossible to detect, even from the most pedestrian stego-systems, while long messages are easily detectable with medieval steganalysers. We introduce a fourth user scenario that we call *storage steganography*, which focuses purely on very long messages. It arises from claims in the literature that paedophiles are using steganography to disguise their imagery. It may be useful for Alice to use steganography to hide the true nature of her data on her own system, even without any plans to communicate the steganograms to Bob. Thus, in the event of a police raid, it will hopefully be impossible for any raiders to identify suspicious material.

Most of the literature focuses on what we can call *one-off steganography*, where Wendy intercepts one steganogram and attempts to determine if it contains a message. In practice, this may not be the most plausible scenario. In an organised conspiracy, the members are likely to communicate regularly over a long period of time. In storage steganography, many steganograms will be needed to fit all the data that Alice wants to hide. Ker (2006) introduced the concept of *batch steganography* to cover this situation. Alice spreads the secret data over a batch of data objects, and Wendy faces a stream of potential steganograms. When she observes a large sample of steganograms, Wendy may have a better probability of getting statistically significant evidence of steganography.

The last question to consider with respect to Alice's viewpoint is message length, and we know that short messages are harder to detect than long ones. It goes without saying that sufficiently short messages are undetectable. Storage steganography would presumably require the embedding of many mega bytes, quite possibly giga bytes, to be useful. The same may be the case for organised crime and terrorist networks if detailed plans, surveillance photos, and maps need to be exchanged via steganography. For comparison, a 512×512 image can store 32 Kb of data using LSB embedding. The capacity for JPEG varies depending on the image texture and compression level, but about 3–4 kb is common for 512×512 images. Thus many images might be needed for steganograms in the above applications, and testing the steganalysis algorithms at 50–100% of capacity may be appropriate.

Operational communications within a subversive team, whether it belongs to secret services or the mob, need not require long messages. It might be a date and a time, say 10 characters or at most 21 bits in an optimised representation. An address is also possible, still 20–30 characters or less. Useful steganographic messages could therefore fit within 0.1% of LSB capacity for pixmap images. Yet, in the steganalysis literature, messages less than 5% of capacity are rarely considered. Any message length is plausible, from a few bytes to as much as the cover image can take.

14.1.2 Wendy's Role

In the prisoners' problem, the very use of steganography is illegal. Furthermore, the prison environment may allow Wendy to impose punishment and withdraw privileges without solid proof. Just suspecting that it takes place may be sufficient for Wendy to block further communication. That is rarely the case in other scenarios. If Wendy is an intelligence agent, and Alice and Bob use steganography to hide a conspiracy, Wendy will usually have to identify the exact conspiracy in order to stop it, and solid proof is required to prosecute.

Very often, basic steganalysis will only be a first step, identifying images which require further scrutiny. Once plausible steganograms have been identified, other investigation techniques may be employed to gather evidence and intelligence. Extended steganalysis and image forensics may be used to gather information about the steganograms. Equally important, supporting evidence and intelligence from other sources may be used.

Even considering just the basic steganalyser, different use cases will lead to very different requirements. We will primarily look at the implications for the desired error probabilities p_{FP} and p_{FN}, but also some specialised areas of steganalysis.

Basic Steganalysis

The background for selecting the trade-off between p_{FP} and p_{FN} is an assessment of risk. We recall the Bayes risk, which we defined as

$$\text{risk} = c_{FP} \cdot p_{FP} \cdot (1 - P(S)) + c_{FN} \cdot p_{FN} \cdot P(S), \qquad (14.1)$$

where c_x is the cost of a classification error and $P(S)$ is the *(a priori)* probability that steganography is being used. We have assumed zero gain from correct classifications, because only the 'net' quantity is relevant for the analysis. The risk is the expected cost Wendy faces as a result of an imperfect steganalyser.

Most importantly, we must note that the risk depends on the probability $P(S)$ of Alice using steganography. Hence, the lower $P(S)$ is, the higher is the risk associated with false positives, and it will pay for Wendy to choose a steganalyser with higher p_{FN} and lower p_{FP}. To make an optimal choice, she needs to have some idea of what $P(S)$ is.

In the prisoners' problem, one might assume that $P(S)$ is large, possibly of the order of 10^{-1}, and we can probably get a long way with state-of-the-art steganalysers with about 5–10% false positives and $p_{FP} \approx p_{FN}$. Storage steganography may have an even higher prior stego probability, as Wendy is unlikely to search Alice's storage without just cause and reason for suspicion.

A web crawler deployed to search the Web for applications of dead-drop steganography is quite another kettle of fish. Most web users do not know what steganography is, let alone use it. Even a steganography user will probably use it only on a fraction of the material posted. Being optimistic, maybe one in a hundred thousand images is a steganogram, i.e. $P(S) \approx 10^{-5}$. Suppose Wendy uses a very good steganalyser with $p_{FP} \approx 1\%$ and $p_{FN} \approx 1\%$. In a sample of a hundred thousand images, she would then expect to find approximately a thousand false positives and one genuine steganogram.

Developing and testing classifiers with low error probabilities is very hard, because it requires an enormous test set. As we discussed in Chapter 3, 4000 images gives us a standard error of roughly 0.0055. In order to halve the standard error, we need four times as many test images. A common rule of thumb in error control coding, is to simulate until 100 error events. Applying this rule as a guideline, we would need about ten million test images to evaluate a steganalyser with an FP rate around $p_{FP} \approx 10^{-5}$.

One could argue that a steganalyser with a high false positive rate would be useful as a first stage of analysis, where the alleged steganograms are subject to a more rigorous but slower and more costly investigation. However, little is found in the literature on how to perform a second stage of steganalysis reliably. The first-stage steganalyser also needs to be accurate enough to give a manageable number of false positives. Wendy has to be able to afford the second-stage investigation of all the false positives. Even that may be hard for a web-crawling state-of-the-art steganalyser. The design of an effective steganalysis system to monitor broadcast channels remains an unsolved problem.

14.1.3 Pooled Steganalysis

Conventional steganalysis tests individual media objects independently. It is the individual file that is the target, rather than Alice as the source of files. Ker (2006) introduced the concept of *batch steganography*, where Alice spreads the secret data across multiple (a batch of) media objects. Wendy, in this scenario, would be faced with a stream of potential steganograms. It is not necessarily interesting to pin point exactly which objects contain secret data. Very often the main question is whether Alice uses steganography. Thus Wendy's objective should be to ascertain if at least one of the files is a steganogram. Traditional steganalysis does not exploit the information we could have by considering the full stream of media objects together.

A natural solution to this problem is called *pooled steganalysis*, where each image in the stream is subject to the same steganalyser, and the output is subject to joint statistical analysis to decide whether the stream as a whole is likely to contain covert information. Ker demonstrated that pooled steganalysis can do significantly better than the naïve application of the constituent algorithm. In particular, he showed that under certain assumptions, when the total message size n increases, the number of covers has to grow as n^2 to keep the steganalyst's success rate constant. This result has subsequently become known as the *square-root law* of steganography, and it has been proved under various cover models. It has also been verified experimentally.

Ker has analysed pooled steganalysis in detail under various assumptions, see for instance Ker (2007a). The square-root law was formulated by Ker (2007b) and verified experimentally by Ker *et al.* (2008). Notably, Ker (2006) showed that Alice's optimal strategy is not necessarily to spread the message across all covers. In some cases it is better to cluster it into one or a few covers.

The square-root law stems from the fact that the standard deviation of a statistic based on an accumulation of N observations grows as $\Theta(\sqrt{N})$. We have seen this with the standard error in Section 3.2 and in the hypothesis test of Section 10.5.2. The proofs that the cover length N has to grow exactly as the square of the message length n are complicated and tedious, but it is quite easy to see that N has to grow faster than linearly in n. If we let N and n grow, while keeping a constant ratio n/N, then Wendy gets a larger statistical sample from the same distribution, and more reliable inference is possible.

In principle, batch steganography refers to one communications stream being split across multiple steganograms sent from Alice to Bob. The analysis may apply to other scenarios as well. If Wendy investigates Alice, and is interested in knowing whether she is using steganography to communicate with at least one of her contacts Bob, Ben or Bill (etc.), pooled steganalysis of everything transmitted by Alice can be useful. According to the square-root law, Alice would need to generate rapidly increasing amounts of innocent data in order to maintain a fixed rate of secret communications, which should make Wendy's task easy over time.

Finally, there is reason to think that the square-root law may tell us something about the importance of image size in the unpooled scenario; that doubling the message length should require a quadrupling of the image size. As far as we know, such a result has neither been proved analytically nor demonstrated empirically, but the argument about the variance of features going as $O(1/\sqrt{N})$ should apply for most features also when N is the number of coefficients or pixels. If this is true, relative message length as a fraction of image size is not a relevant statistic.

14.1.4 Quantitative Steganalysis

Quantitative steganalysis aims to estimate the length of the hidden message. Both the approaches and the motivations vary. Where simple steganalysis is a classification problem, quantitative steganalysis is an estimation problem. Alice has embedded a message of q bits, where q may or may not be zero. We want to estimate q. Knowledge of q may allow Wendy to assess the seriousness of Alice and Bob's activities, and can thereby give useful intelligence.

We can formulate quantitative steganalysis as a regression problem, and as discussed in Section 12.3, both statistical and learning algorithms exist to solve it. Any feature vector which can be used with a learning classifier for basic steganalysis can in principle be reapplied with a regression model for quantitative steganalysis.

Quantitative steganalysis is not always an end in its own right. Sometimes we want to develop a statistical model to identify good features for classification, and we find an estimator $\hat{q}$ for q as a natural component in the model. Quite commonly $\hat{q}$ is abused as a measure of confidence; the larger $\hat{q}$ is the more confidently we conclude that the image contains a hidden message. Even though this idea can be justified qualitatively, there is no clear quantitative relationship. We cannot use $\hat{q}$ directly to quantify the level of confidence or significance.

Applying a quantitative steganalyser to the basic steganalysis problem is possible, but it is not as straightforward as one might think. In fact, it is very similar to using any other feature for classification. We have to select a threshold τ and assume a steganogram for $\hat{q} > \tau$ and a clean image for $\hat{q} < \tau$. The threshold must be selected so that the error rates $P(\hat{q} > \tau \mid q = 0)$ and $P(\hat{q} < \tau \mid q \gg 0)$ are acceptable. Optimising the threshold τ is exactly what a classification algorithm aims to do.

Sometimes, one may want to assess $\hat{q}$ without explicitly selecting τ, merely making a subjective distinction of whether $\hat{q} \approx 0$ or $\hat{q} \gg 0$. This can give an implicit and subjective assessment of the level of confidence. A numeric measure of confidence is very useful, but there are better alternatives which allow an objective assessment of confidence. A good candidate is a Bayesian estimate for the posterior probability $P(\text{stego} \mid I)$. The p-value from a hypothesis test also offers some objectivity. In libSVM, for example, such probability estimates can be obtained using the $-b$ 1 option.

14.2 Images and Training Sets

As important for steganalysis as the selection of features, is the choice of training set. In the last few years, steganalysts have become increasingly aware of the so-called *cover-source mismatch* problem. A steganalyser trained on images based on one cover source may not perform nearly as well when presented with images from a different source.

A closely related question is the one asked by Alice. What cover source will make the steganograms most difficult to detect? Wendy should try to use a cover source resembling that used by Alice. Without knowing exactly what Alice knows, it is useful to know how to identify covers which make detection hard. In this chapter we will consider some properties of covers and cover sources that affect detection probability.

14.2.1 Choosing the Cover Source

It is straight forward to compare given cover sources, and a number of authors have indeed compared classes of covers, focusing on different characteristics. For each class of cover images, we can train and test a steganalyser; and compare the performance for the same feature vector and classifier algorithm but for different cover sources. If the steganalyser has poor performance for a given source, that means this particular source is well suited from Alice's point of view.

This only lets us compare sources. Searching for the ideal cover source remains hard because of the vast number of possibilities. Furthermore, the analysis cannot be expected to be consistent across different steganalysers, and we would therefore have to repeat the comparison for different steganalysers, in much the same way as a comparison of steganalysers must be repeated for different cover sources.

Image Sources

From capturing to its use in steganography or steganalysis, an image undergoes a number of processes. First is the physical capturing device, with its lenses and image sensor. Second, the image may be subject to in-camera image processing and compression, and finally, it may be subject to one or more rounds of post-processing and recompression. Post-processing may include colour, brightness and contrast correction or other tools for quality enhancement. It may also include downsampling, cropping or recompression to adapt it for publication in particular fora.

The motive and texture of the image can affect the results. Light conditions will have obvious and significant effects on image quality, and there may also be differences between outdoor/indoor, flash, lamp light and daylight, and between people, landscapes and animals.

Many of the differences between different image sources can be explained in terms of noise. Many of the known feature vectors are designed to detect the high-frequency noise caused by embedding, but obviously the steganalyser cannot distinguish between embedding distortion and naturally occurring noise, and clean images have widely varying levels of natural noise. Highly compressed images, where the high-frequency noise has been lost in compression, will offer less camouflage to hide the embedded message. Contrarily, images with a lot of texture will offer good camouflage for a hidden message. Different capturing devices may cause different levels of high-frequency noise depending on their quality.

Authors in steganalysis typically draw their images from published collections such as UCID (Schaefer and Stich, 2004), Corel, Greenspun (2011) or NRCS (Service, 2009). This will usually mean high-quality images, but we usually do not know if they have been subject to post-processing. Some also use own-captured images, and again, as readers we do not know the characteristics of these images. A number of image databases have been published specifically for research in steganography and watermarking, and these tend to be well-documented. Most notable are the two 10 000 image databases used for the BOWS (Ecrypt Network of Excellence, 2008) and BOSS (Filler *et al.*, 2011b) contests, but also one by Petitcolas (2009).

In an extensive performance analysis of steganalysers based on Farid-72, FRI-23 and BSM-*xx*, Kharrazi *et al.* (2006b) used a large number of images acquired with a web crawler. This gave them a more diverse set of images than previous evaluations. They separated the images into high, medium and low compression quality, and trained and tested the steganalysers on each class separately. Unsurprisingly, the low-quality images gave the most detectable steganograms.

The capturing device is known to have a significant effect on the detectability. Wahab (2011) has tested this for Farid-72 and FRI-23, where the steganalyser was trained and tested on images from different capturing devices. It was found that FRI-23 was remarkably robust against changes in image source, whereas Farid-72 was significantly less accurate when it was trained and tested on different image sources.

The question of capturing device can be very specific. Goljan *et al.* (2006) showed that detection performance may improve by training the classifier on images from the exact same camera, compared with training on images from another camera of the same make and model. This may be surprising at first glance, but should not be. Microscopic devices, such as the image sensor, will invariably have defects which are individual to each unit produced. Adequate performance is achieved through redundancy, so that a certain number of defects can be tolerated. Defects in the image sensor cause a noise pattern which is visually imperceptible, but which may be picked up statistically by a steganalyser or similar system.

Another important image characteristic, which has been analysed by Liu *et al.* (2008), is that of texture. They considered the diagonal sub-band of the Haar wavelet decomposition in order to quantify the texturing. The distribution of coefficient values is modelled using a generalised Gaussian distribution (GGD), which is defined as

$$p(x; \sigma, \beta) = \frac{\beta}{2\sigma \Gamma(1/\beta)} \exp\left(\frac{-|x|}{\sigma}\right)^{\beta},$$

where $\Gamma(z) = \int_0^\infty e^{-t} t^{z-1} dt$ for $z > 0$ is the Gamma function. The σ parameter is just the standard deviation, but the so-called shape parameter β of the diagonal Haar band is a measure of the texture complexity of the pixmap image. Liu *et al.* showed that steganalysers with an accuracy of more than 90% for images with little texture ($\beta < 0.4$) could be as bad as $< 55\%$ for highly textured images with $\beta > 1$.

Choosing Cover for Embedding

The selection of a good cover is part of Alice's strategy and has long been recognised as an important problem even if it remains only partially solved. Lately, some authors have started to devise methods to filter the cover source to select individual images that are particularly well suited.

Filter methods may be general or message-specific. A general method would determine whether a given cover is suitable for messages of a particular length, whereas a message-specific method would check if a given steganogram, with a specific message embedded, is likely to go undetected.

In one of the earlier works on the topic, Kharrazi *et al.* (2006a) analysed a number of heuristics for cover selection. General methods can be as simple as counting the number of changeable coefficients. Embedding in a cover with many changeable coefficients means that the relative number of changes will be small. For LSB embedding this obviously corresponds to image size, but for JPEG embedding, zero coefficients are typically ignored, so highly textured or high-quality images will tend to give less detectable steganograms.

Sajedi and Jamzad (2009) proposed a general filter method which is considerably more expensive in computational terms. They estimated embedding capacity by trial and error, by embedding different length messages and checking the detectability in common steganalysers. The capacity can then be defined as the length of the longest undetectable message. The images and their capacity can then be stored in a database. When a given message is going to be sent, the cover is selected randomly from those images with sufficient capacity.

14.2.2 The Training Scenario

By now, it should be clear that all universal steganalysers must be very carefully trained for the particular application. A steganalyser is designed to detect artifacts in the images. It will pick up on any artifact which statistically separates the classes in the training set, whether or not those artifacts are typical for the two classes in the global population. These artifacts may or may not be typical for steganography, and they are unlikely to be unique to steganography. For instance, a steganalyser which is sensitive to the high-frequency noise caused by embedding will make many false positives when it is presented with clean images with a higher level of naturally occurring noise.

The design of the training scenario is part of the steganalyst strategy. Obviously, it requires an understanding of Alice's choice of images as discussed in the previous section. Furthermore, the cover source favoured by Alice the steganographer is not necessarily related to the source of clean images transmitted by the innocent Alice. Hence there are potentially two different image distributions that must be understood. This issue may be too hard to tackle though; all known authors use the same source for clean images and cover images. But we will look into a couple of other concerns that we need to be aware of regarding the preparation of images.

Image Pre-processing

Many of the common features in steganalysis may depend on irrelevant attributes such as the image size. It is therefore customary to pre-process the images to make them as comparable as possible.

The most obvious example of size-dependent features are those derived from the histogram. Very often the absolute histogram is used, and a larger image will have larger histogram values. This can be rectified by using the relative histogram, which is the absolute histogram divided by the total number of coefficients. Other features may depend on size in a less obvious way.

The most common approach to dealing with this problem is to crop all images to one pre-defined size. Scaling has been used by some authors, but is generally considered to be more likely to introduce its own artifacts than cropping is. This approach works well for training and testing of the steganalyser, but it does not tell us how to deal with off-size images under suspicion. We can crop the intercepted image prior to classification, but then we discard information, and what happens if the secret message is hidden close to the edges and we crop away all of it? We could divide the image into several, possibly overlapping, regions and steganalyse each of them separately, but then the problem would essentially become one of detecting batch steganography and we should design the steganalyser accordingly.

For a production system, all the features should be normalised to be independent of the image size. This is not necessarily difficult, but every individual feature must be assessed and normalised as appropriate, and few authors have considered it to date.

Double Compression

Of all the irrelevant artifacts which can fool a steganalyser, double compression is probably the strongest, and certainly the most well-known one. It is the result of a bug in the common implementations of algorithms like F5 and JSteg. When the software is given a JPEG image as input, this is decompressed using common library functions, and the quantisation tables discarded. The image is then recompressed using default quantisation tables, possibly with a user-defined quality factor, and the embedding is performed during recompression.

The result is that the steganograms have been compressed twice, typically with different quality factors. This double compression leaves artifacts which are very easy to detect. Early works in steganalysis would train the steganalyser to distinguish between single-compressed natural images and double-compressed steganograms. What are we then detecting? Double compression or steganographic embedding?

This training scenario will fail both if Alice has a different implementation of F5, or she uses uncompressed images as input; and it will fail if doubly compressed but clean images occur on the channel. The first case will lead to a high false negative probability, and the second to many false positives. The problem was first pointed out by Kharrazi *et al*. (2006b), who compared a steganalyser with a training set where both natural images and steganograms were double compressed. This training scenario gave much lower accuracies than the traditional training scenario. Later, Wahab (2011) made similar tests with a training set where both the natural images and steganograms were compressed once only, and found similar results.

Double compression is not uncommon. Images captured and stored with high-quality compression will often have to be recompressed at a lower quality to allow distribution on web pages or in email. It is also common that images are converted between image formats, including uncompressed formats, as they are shared between friends, and this may also lead to recompression at a higher quality factor than previous compressions, as the user will not know how the image previously compressed. Thus it is clear that evidence of double compression is no good indication of steganography, and the steganalyser has to be able to distinguish between steganograms and clean images whether they are singly or doubly compressed.

It is obvious that we need to train the steganalyser on every embedding algorithm we expect steganographers to use. The double-compression problem tells us that we may have to train it for every conceivable *implementation* of the algorithm as well. In addition to the well-studied double-compression bug, there may be other implementation artifacts which have not yet been discovered.

Embedding and Message Length

Each steganogram in the training set has to be embedded with some specific message using a specific algorithm. We usually train the steganalyser for a specific embedding algorithm, and the significance of this choice should be obvious.

The choice of message can also affect the steganalysis. The visual steganalysis discussed in Section 5.1 showed us how different types of messages leave different artifacts in the LSB plane. In general, higher-order features studying relationships between different bits or pixels will respond very differently to random message data and more structured data such as text. The most common choice, especially in recent years, is to use random messages for training and testing, and there are many good reasons for this.

In real life, random-looking messages are the result of compressing and/or encrypting the data. Both compression and encryption are designed to produce random-looking data. Compression because it removes any redundancy which would cause the statistical structure, and encryption because random-looking data gives the least information for the eavesdropper. Compression will reduce the number of embedding changes and thereby make the embedding less detectable. Alice may also want to encrypt the message to protect the contents even if the existence were to be detected. Hence, random-looking messages are very plausible.

If Alice were to go for non-random-looking data she should, as we saw in the discussion of Bayes' classifiers, aim for a probability distribution resembling that of the cover image. The easiest way to achieve this is to use a random-looking message and embed it in random-looking data. Otherwise, it is very hard to find a message source and a cover source with matching distributions. Although some may exist, there are no known examples where a random-looking message is easier to detect than a non-random one. Again, Alice has the best chance by using random-looking messages.

The next question for the training phase is what message lengths to consider. Commonly, new classifiers are trained and tested for a number of message lengths, specified as a fraction of capacity, typically ranging from 5% to 100% . This has a number of problems. Although capacity is well-defined for LSB embedding, as one bit per pixel, it is ill-defined for many other algorithms. In JPEG steganalysis, one often specifies the message length in bits per non-zero coefficients (bpc). This overestimates capacity for JSteg and Outguess 0.1, which cannot use coefficients equal to one either. In F4 and F5 the capacity in bits will depend on the contents of the message, because re-embedding occurs as the result of specific bit values.

All of these measures relate, even if loosely, to the physical capacity of the cover image, and are unrelated to the detectability. The square-root law indicates that even if 5% of messages are detectable for a given size of cover image, they may not be detectable if sufficiently small images are used. Detectability depends both on cover size and message length, not just on the ratio between them.

In this book, we have tended to use a constant message length in bytes or in bits for all images. For LSB embedding with a fixed image size in the spatial domain, this is of course equivalent to fixing a relative message length. For JPEG, in contrast, it will be different because the capacity will depend on image texture as well as image size. The choice to use constant length in bits is motivated by the practical requirements seen by Alice. She will have a specific message which she needs to send to Bob. The message length is determined only by Alice's usage scenario and not at all by the cover image. This approach, however, ignores cover selection. If Alice's uses cover selection, many of the cover's in the training and test sets would be ineligible as covers for the particular message length.

Capacity and Message Length for F5

When designing a training scenario, it is useful to consider what parameters the stego-system has been designed for. This is especially true when steganalysis is used to evaluate the stego-system. Far too often, stego-systems are evaluated and criticised under conditions they were never intended for. Take F5 as an example. Using F5 at maximum capacity would mean that every non-zero coefficient is needed for the payload, and matrix coding cannot be used to reduce the number of embedding changes. Thus F5 degenerates to F4.

Briffa *et al.* (2009b) report some tests of the capacity of F5. The matrix code is an $[2^k - 1, 2^k - 1 - k, 3]$ Hamming code, which means that k bits are embedded using $2^k - 1$ changeable coefficients. According to Briffa *et al.*, embedding at $k = 1$ typically hides 1.4 message bits per modified coefficients. At $k = 4$ we almost always get two message bits per change, typically 2.1–2.5. At $k = 7$, which is the maximum k in Westfeld's implementation, we can typically embed 3.5–4 bits per change.

In order to reduce the distortion by 50% compared to F4, the minimum value is $k = 5$, which gives a code rate of between 1/6 and 1/7. Because of the re-embeddings where zeros are created, the embedding rate in bpc will be considerably lower than the code rate. At $k = 7$ we have a code rate of 0.055, while the empirical embedding capacity according to Briffa *et al.* is about 0.008 bpc on average. In other words, in order to get a significant benefit from matrix coding, F5 should only be used at embedding rates of a few percent of the number of non-zero coefficients. In the literature we often see 5% as the low end of the range of message lengths considered, but for F5 it should rather have been the high end.

The Size of the Training Set

The required size of the training set is probably the most obvious question regarding the training scenario. We have left it till the end because it is also the hardest. Miche *et al.* (2007) did suggest an approach to determine a good training set size for a given cover source. Starting with a total training set of 8 000 images, they ran a Monte Carlo simulation repeatedly, selecting a smaller training set randomly. The classifier was trained on the smaller set and tested on a separate set of 5000 images. Testing different training set sizes, from 200 to 4000, they found that the accuracy converged and little was gained from using more than 2000 images for testing.

There is no reason to expect that the result of Miche *et al.* generalises. The more diverse the image base is, the larger the training set will have to be in order to be representative. It is, however, possible to use their methodology to determine a sufficient training set for other scenarios. Where one has access to a large image base, this may be a useful exercise.

14.2.3 The Steganalytic Game

The steganography/steganalysis problem can be phrased as a two-player game, where Alice and Wendy are the players. Alice transmits one or more files, with or without steganographic contents. Wendy attempts to guess whether there is a steganographic content or not. If Wendy guesses correctly she wins, otherwise she loses. When the game is complete, each player receives a payoff, which is usually normalised for a zero-sum game where the winner receives an amount x and the loser receives $-x$. The payoff is essentially the same quantity as the costs c_{FP} and c_{FN} which we used to define risk.

Considering practical applications of steganalysis, the game is exceedingly complex. Both Alice and Wendy have a number of strategic choices to make. Alice can not only choose whether or not to use steganography, she can also choose

- stego-algorithm,
- cover image,
- pre-processing of the cover image,
- message length,
- and possibly transmission channel.

Similarly, Wendy can choose

- feature vector,
- classification algorithm,
- training set(s), including
  — cover source,
  — stego-algorithm(s),
  — message length(s); and
- decision threshold (trade-off between p_{FP} and p_{FN}).

Clearly, the probability of success of Wendy's strategy will depend on Alice's strategy and vice versa. Wendy's optimal strategy may no longer be optimal if Alice changes hers.

Game theory is a well-established field within economics and mathematics, and has found applications in many other areas, including engineering problems such as power control in cellular networks. In theory, it allows us to calculate optimal strategies for the two players analytically. However, the strategy options we have listed above are far too many to allow an analytic solution. Instead, the formulated game highlights the complexity of the problem.

A key characteristic of any game is who moves first. Does one player know the strategy of the other player before she makes her own move? If Alice knows Wendy's strategy, she would of course test tentative steganograms against Wendy's strategy and reject any which are detectable. This may be the case if Wendy is part of a large-scale public operation, as it is then difficult to keep her methods secret. Similarly, if Wendy knows Alice's strategy, she would tune her steganalyser to that strategy. Although it may be unlikely that Wendy knows Alice's strategy in full, she might have partial information from other investigations, such as knowing that Alice has a particular piece of stego software installed.

Looking at Wendy's problem in general, it is reasonable to expect that Wendy and Alice move simultaneously and have to choose a strategy with no knowledge of the other player.

14.3 Composite Classifier Systems

A simple steganalyser as discussed so far consists of a single instance of the classifier algorithm. It is trained once, resulting in one model instance, which is later used for classification. The simple steganalyser is tuned for a particular training set, and may have poor performance on images which are very different from the training set. Very often, it is useful to combine several steganalysers or classifiers into a more complex steganalysis system.

We will start with a discussion of fusion, where multiple constituent classifiers are designed to solve the same classification problem, and then combined into a single classifier which will hopefully be more accurate. Afterwards, we will look at other composite steganalysis systems, where the classification problem is split between different classifiers, both steganalysers and other classifiers.

14.3.1 Fusion

Fusion refers to several standard techniques in machine learning. Very often, good classification results can be obtained by combining a number of mediocre classifiers. The simplest form of fusion is feature-level fusion, where multiple feature vectors are fused together to form a new, higher-dimensional feature vector. The fused feature vector can be used exactly like the constituent feature vectors, and we discussed examples of this in Chapter 13 without thinking of it as a new technique. The main drawback of feature-level fusion is the curse of dimensionality, and it is often necessary to combine it with feature selection to get reasonable performance. Other than the increase in dimensionality there is no change to the classification problem, and we will typically use the same algorithms as we do in the unfused scenario.

It is also possible to fuse classifiers based on the output of the constituent classifiers. Figure 14.1 shows how fusion operates at different layers with four different

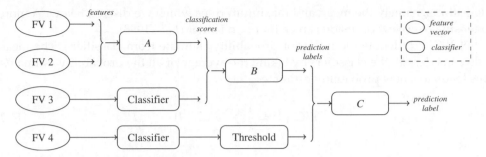

Figure 14.1 The three types of fused classifiers: feature-level (*A*), heuristic-level (*B*), and classifier-level (*C*)

constituent feature vectors FV1–4. Classifier *A* uses feature-level fusion as discussed before; the two feature vectors FV1 and FV2 jointly form the input for *A*. Classifier *B* uses heuristic-level or measurement-level fusion, where the classification scores of the constituent classifiers form the input. Finally, classifiers can be fused at the classifier level or abstract level, as classifier *C*. Where heuristic-level classifiers use the soft information from the constituent classifiers, a classifier-level classifier uses the discrete label predictions after the classification scores have been thresholded.

Both heuristic-level and classifier-level fusion avoid the curse of dimensionality unless a truly impressive number of constituent classifiers is used, but there is also another advantage over feature-level fusion. When we augment an existing classifier with new features, feature-level fusion requires complete retraining (and possibly new feature selection), which can be expensive when large data sets are involved. When fusion is based on classifier output, we can simply train a new classifier using the new features and add it to the fused classifier without retraining the other constituent classifiers. Since the fused classifier works with fewer features, and often uses a simpler algorithm, it will be much faster to train.

There is a wide range of possible algorithms which may be suitable for the fused classifier. One possibility is to view the input classification scores like any other feature vector and use a generic algorithm like SVM, FLD or neural networks, just as in the unfused case. Thus no new theory is necessary to handle fusion. This approach is also simple because we do not need any semantic interpretation of the classification scores. The constituent classifiers do not even have to produce scores on the same scale; instead we trust the learning algorithm to figure it out. More commonly, perhaps, the heuristic-level fused classification problem is solved by much simpler algorithms, exploiting the fact that we usually have much more knowledge about the probability distribution of the classifier output than we have for the original features.

Many classifiers, including Bayesian ones, output *a posteriori* probabilities $\hat{P}(C_I \mid I)$ that a given image I belongs to class i, and it is usually possible to get such probability estimates by statistical analysis of the classifier output. Given probability estimates for each class, it is possible to base the fused classifier on analysis instead of machine learning. In the context of steganalysis, Kharrazi *et al.* (2006c) tested two such fusion algorithms, namely the mean and maximum rules which we discuss below. A more extensive overview of fusion can be found in Kittler *et al.* (1998).

Let $\hat{P}_j(C_I \mid I)$ denote the posterior probability estimate from classifier j. The mean rule considers all the classifiers by taking the average of all the probability estimates; this leads to a new probability estimate

$$\hat{P}(C_j \mid I) = \frac{1}{k} \sum_i \hat{P}_i(C_j \mid I), \tag{14.2}$$

and the predicted class $\hat{c}$ is the one maximising $\hat{P}(C_j \mid I)$. The mean rule is based on the common and general idea of averaging multiple measurements to get one estimate with lower variance, making it a natural and reasonable classification rule.

The maximum rule will trust the classifier which gives the most confident decision, i.e. the one with the highest posterior probability for any class. The predicted class is given as

$$\hat{c} = \arg \max_{X \in \{C,S\}} \max_i P_i(C \mid I).$$

The maximum rule is used in the one-against-one and one-against-all schemes used to design multi-class classifiers from binary ones. It is a logical criterion when some of the classifiers may be irrelevant for the object at hand, for instance because the object is drawn from a class which was not present in the training set. Using the mean rule, all of the irrelevant classifiers would add noise to the fused probability estimate, which is worth avoiding. By a similar argument, one could also expect that the maximum rule will be preferable when fusing classifiers trained on different cover sources.

In classifier-level fusion, the input is a vector of discrete variables $(\hat{y}_1, \ldots, \hat{y}_k)$ where each $\hat{y}_i$ is a predicted class label rather than the usual floating point feature vectors. Therefore standard classification algorithms are rarely used. The most common algorithm is majority-vote, where the fused classifier returns the class label predicted by the largest number of constituent classifiers. Compared to heuristic-level fusion, classifier-level fusion has the advantage of using the constituent classifiers purely as black boxes, and thus we can use classifier implementations which do not provide ready access to soft information. On the contrary, if soft information is available, classifier-level fusion will discard information which could be very useful for the solution.

14.3.2 A Multi-layer Classifier for JPEG

Multi-layer steganalysis can be very different from fusion. Fusion subjects the suspect image to every constituent classifier in parallel and combines the output. In multi-layer steganalysis, the classifiers are organised sequentially and the outcome of one constituent classifier decides which classifier will be used next.

A good example of a multi-layer steganalysis system is that of Pevný and Fridrich (2008) for JPEG steganalysis. They considered both the problem of single and double compression and the issue of varying quality factor in steganalysis. To illustrate the idea, we consider quality factors first. We know that the statistical properties of JPEG images are very different depending on the quality factor. The core idea is to use one classifier (or estimator) to estimate the quality factor, and then use a steganalyser which has been trained particularly for this quality factor only. Because the steganalyser is specialised for a limited class of cover images, it can be more accurate than a more general one.

The design of multi-layer classifiers is non-trivial, and may not always work. Obviously we introduce a risk of misclassification in the first stage, which would then result in an inappropriate classifier being used in the second stage. In order to succeed, the gain in accuracy from the specialised steganalyser must outweigh the risk of misclassification in the first stage. The quality factor is a good criterion for the first classifier, because the statistical artifacts are fairly well understood. It is probably easier to estimate the quality factor than it is to steganalyse.

Remark 14.1 *The feature-based classifier may not be the best approach to estimate the quality factor. When an image is distributed as JPEG, one can accurately calculate the quality factor from the quantisation matrix, with no need for statistical analysis or machine learning. A feature-based classifier would only make sense when previously compressed images are distributed as pixmap images.*

Another problem which can be tackled by a multi-layer classifier is that of double and single compression of JPEG steganograms. Just as they distinguished between compression levels, Pevný and Fridrich (2008) proposed to resolve the double-compression problem by first using a classifier to distinguish between singly and doubly compressed images, and subsequently running a steganalyser specialised for doubly or singly compressed images as appropriate.

14.3.3 Benefits of Composite Classifiers

Heuristic-level fusion is most often described as a fusion of constituent classifiers using different feature vectors. Implicitly one may assume that all the classifiers are trained with the same training sets. However, the potential uses do not stop there.

Fusion is a way to reduce the complexity of the classifiers, by reducing either the number of features or the number of training vectors seen by each constituent classifier. This can speed up the training phase or improve the accuracy, or both. It can also make it possible to analyse part of the model, by studying the statistics of each constituent classifier. When we discussed the application of SPAM features for colour images, it was suggested to train one classifier per colour channel, and then use heuristic- or abstract-level fusion to combine the three. Thus, each classifier will see only one-third of the features, and this reduces the risk of the curse of dimensionality and may speed up training.

The multi-layer classifier described in the previous section uses constituent classifiers with the same feature vector but different training sets. Kharrazi *et al.* (2006c) proposed using the same idea with fusion, training different instances of the classifier against different embedding algorithms. When new stego-systems are discovered, a new constituent classifier can be trained for the new algorithm and added to the fused classifier. This is essentially the one-against-one scheme that we used to construct a multi-class SVM in Section 11.4, except that we do not necessarily distinguish between the different classes of steganograms. Thus, it may be sufficient to have one binary steganalyser per stego-system, comparing steganograms from one system against clean images, without any need for steganalysers comparing steganograms from two different stego-systems.

On a similar note, we could use constituent classifiers specialised for particular cover sources. The general problem we face in steganalysis is that the variability within the class of clean images is large compared to the difference between the steganograms and clean images. This is the case whether we compare feature vectors or classification heuristics. Multi-layer steganalysis is useful when we can isolate classes of cover images with a lower variance within the class. When the steganalyser is specialised by training on a narrower set of cover images it can achieve higher accuracy, and this can form the basis for a multi-layer steganalyser. There are many options which have not been reported in the literature.

14.4 Summary

The evolution of steganography and steganalysis is often described as a race between steganalysts and steganographers. Every new stego-system prompts the development of a new and better steganalysis algorithm to break it; and vice versa. This chapter illustrates that it is not that simple.

New algorithms have been tested under laboratory conditions, with one or a few cover sources, and the detection accuracies normally end up in the range of 70–90%. Even though some authors test many sets of parameters, the selection is still sparse within the universe of possible use cases. Very little is known about field conditions, and new methods are currently emerging to let Alice select particularly favourable cover images, making the steganalyst laboratory conditions increasingly irrelevant. Removing the focus from feature vectors and embedding algorithms, the race between Alice and Wendy could take on a whole new pace . . .

The parameters in the recent BOSS contest (Bas *et al.* 2011) are typical for empirical discussions of steganalysis, and may serve as an illustration without any intended criticism of the BOSS project, which has resulted in a few very useful additions to the literature. The data set for the HUGO stego-system used steganograms at 40% of capacity, and the winner achieved 80.3% accuracy. From a theoretical point of view, this has led to the conclusion that HUGO is no longer undetectable. Yet two questions are left unanswered here, as for many other stego-systems and steganalysers:

1. Is 80.3% accuracy sufficient for Wendy's real-world application?
2. Does Alice really need to embed at 40% of capacity in reality, or could she run her conspiracy with steganograms at a much smaller fraction of capacity?

Such questions should be asked more often.

15

Future of the Field

Reviewing the state of the art in steganalysis, it is reasonable to ask what future the area has. It may be a future within science and research, or it could be as an applied and practical art. It could be both or neither.

Steganalysis no longer enjoys the broad and popular support it did in the first couple of years of the millennium. As we discussed in early chapters, the hard evidence of real steganography users is missing from the public domain. Nevertheless, there are a number of experts, both in industry and academia, who maintain that steganography and steganalysis are important fields with potential for increasing future use. In 2010 we saw the BOSS contest, *Break Our Steganographic Scheme* (Bas *et al.*, 2011), where a number of groups competed to get the record accuracy for a steganalyser against a particular steganographic system. The subsequent issue of the *Information Hiding Workshop* (Filler *et al.*, 2011a) contained a large number of solid contributions within the area of steganography and steganalysis not limited to BOSS-related work. Is this showing the return of steganography as a hot topic?

15.1 Image Forensics

A field closely related to image steganalysis is image forensics, and as the steganography hype at the start of the century started to cool, several distinguished individuals and groups in steganalysis changed focus. We can view image steganalysis as a special case of image forensics, which aims to discern as much information as possible about the source and history of an image. While image steganalysis is only interested in a single aspect of the image, namely covert information, image forensics covers a wide range of questions, including the following well-known ones:

Camera source identification aiming to classify images according to the camera or capturing device which produced them.
Printer source identification aiming to identify the device which produced a given hard copy.

Machine Learning in Image Steganalysis, First Edition. Hans Georg Schaathun.
© 2012 John Wiley & Sons, Ltd. Published 2012 by John Wiley & Sons, Ltd.

Post-processing analysis aiming to classify images according to what kind of image processing they have undergone, including compression, colour correction, brightness adjustment, etc.

Synthetic image classification aiming to distinguish between natural photographs and computer-generated images.

Tamper detection aiming to detect if an image has been modified ('Photoshopped') after capture.

All the different areas of image forensics have important practical uses. Tamper detection might be the one with the most obvious and wide-ranging applications. In recent years, we have seen a number of tampered images published in the international press, where the tampered image was intended to tell a more serious or scary story than the original image would have. In several cases, news agencies have been forced to recall imagery or let staff go (e.g. BBC News, 2006). The motivation for image tampering may vary, from a freelance photographer yearning for a reputation, to the propaganda of a national government. Either way, news editors will be looking for assurance that images received are genuine and honest, and image forensics is one possible source of such assurance. A press agency or newspaper may receive thousands of images daily from the general public, so the task is not trivial.

Synthetic image classification has received attention because of a number of US court cases involving child pornography. Whereas real photography displaying child pornography is evidence of child abuse which is clearly illegal, not only in the USA but in most countries, it is argued that computer graphics displaying pornographic scenes where no real child is actually involved, would be legal under free speech. Hence, the prosecutor may have to prove that the images presented as evidence come from a real camera, and have not been synthetically generated. Image forensics using machine learning could solve this problem.

Illicit photographs or prints, can possibly be traced back to the offender by using printer and camera identification. A printout may be evidence of industrial espionage, and identifying the printer is one step towards identifying the user. Similarly, illegal imagery can be traced to a camera as a step towards identifying the photographer. The application is very similar to more traditional forensic techniques, used for instance to identify mechanical type writers.

Other areas of image forensics concern more general investigation of the history of a given image. Without directly answering any hard and obvious questions, the applications of image forensics may be subtler. It is clear, though, that knowledge of the history of the image establishes a context which can add credibility to the image and the story behind it. It is very difficult to forge a context for a piece of evidence, and the person genuinely presenting an image should be able to give details about its origin. Thus forensic techniques may validate or invalidate the image as evidence. For instance, a source claiming to have taken a given photograph should normally be able to produce the camera used. If we can check whether the photograph matches the alleged camera, we can validate the claim.

There are many successful examples of steganalysers being retrained to solve other forensic problems. Lyu and Farid's 72-D feature vector is a classic, and it was used to distinguish between real and synthetic photographs in Lyu and Farid (2005). Another classic, Avcibas *et al.* (2004), used image quality measures for tamper detection (Chupeau, 2009). More recently, Wahab (2011) showed that the JPEG features discussed in Chapter 8 can discriminate between different iPhone cameras, and he achieved particularly good results with the Markov-based features and with conditional probabilities. Even if steganalytic feature vectors do not necessarily provide optimised forensic techniques, they make a good background for understanding image forensics in general.

15.2 Conclusions and Notes

Steganalysis has come a long way over the last 10 years, with a good portfolio of attacks. All known stego-systems have been declared broken, typically meaning that steganalysers have been demonstrated with 80–90% accuracy or more.

The fundamental short coming of the current state of steganalysis research is the missing understanding of real-world use cases. It remains an open question if 80–90% accuracy is good or bad for a steganalyser. Similarly, it is an open question what embedding capacity one should expect from a good stego-system. As long as no means to answer these questions is found, the design targets for new systems will remain arbitrary, and quite likely guided by feasibility rather than usefulness.

A second fundamental question for further research is a statistical model for clean images that could allow us systematically to predict detectability of known steganographic distortion patterns. Again, this would be an enormous task, but there are many related open questions which may be more feasible in the immediate term, such as

Cover selection improved methods for feature selection to allow Alice to outwit current steganalysers.

Steganalysis for adverse cover sources some cover sources are already known to be difficult to steganalyse, even without cover source mismatch. The design of steganalysers targeting such sources would be useful, if possible.

Such questions are likely so see increasing attention over the next couple of years.

In spite of all stego-systems being reportedly broken, there is a good case to claim that the steganographer has the upper hand. Steganalysis as we know it is focused entirely on steganography by modification. Yet, it is well known that there are many options for Alice to think outside the box, while steganalysis remains inside it. Steganography by cover selection and cover synthesis is an example of this. Even though we know very little about how to design stego-systems using cover selection or synthesis, even less is known about how to steganalyse them. Just as in classical times, Alice's best shot may very well be to devise a simple *ad hoc* stego-system relying on secrecy of the algorithm. As long as she keeps a low profile and steganograms remain rare, Wendy does not stand much of a chance of detecting them.

The theory of Bayesian inference illustrates this point very well. If the prior probability of observing a steganogram is (say) one in a thousand, a steganalyser with 10% error probability will not have any significant effect on the posterior probability.

We can expect very interesting steganalysis research over the next couple of years. It would be arrogant to try to predict exactly what questions will be asked and which will be answered, but it should be safe to say that they will be different from the questions dominating the last 10 years of research.

Bibliography

Adee, S. 2010. Russian spies thwarted by old technology? *IEEE Spectrum*. http://spectrum
.ieee.org/tech-talk/computing/networks/russian-spies-thwarted-by-old-technology.

Ahmad, I. and Lin, P.E. 1976. A nonparametric estimation of the entropy for absolutely continuous distributions. *IEEE Transactions on Information Theory* **22**, 372–375.

Aizerman, M., Braverman, E. and Rozonoer, L. 1964. Theoretical foundations of the potential function method in pattern recognition learning. *Automation and Remote Control* **25**, 821–837.

Asadi, N., Jamzad, M. and Sajedi, H. 2008. Improvements of image-steganalysis using boosted combinatorial classifiers and Gaussian high pass filtering. IIHMSP 2008 International Conference on Intelligent Information Hiding and Multimedia Signal Processing, pp. 1508–1511.

Avcibaş, I., Sankur, B. and Sayood, K. 2002. Statistical evaluation of image quality measures. *Journal of Electronic Imaging* **11**(2), 206–223.

Avcibaş, I., Memon, N. and Sankur, B. 2003. Steganalysis using image quality metrics. *IEEE Transactions on Image Processing* **12**(2), 221+.

Avcibaş, I., Bayram, S., Memon, N., Ramkumar, M. and Sankur, B. 2004. A classifier design for detecting image manipulations. ICIP 2004 International Conference on Image Processing, vol. 4, pp. 2645–2648.

Avcibaş, I., Kharrazi, M., Memon, N. and Sankur, B. 2005. Image steganalysis with binary similarity measures. *EURASIP Journal on Applied Signal Processing* **17**, 2749–2757.

Backes, M. and Cachin, C. 2005. Public-key steganography with active attacks. *Theory of Cryptography*, vol. 3378 of Lecture Notes in Computer Science. Springer-Verlag.

Bacon, F. 1605. *Of the Proficience and Advancement of Learning Divine and Humane.*

Bagnall, R. 2003. Reversing the steganography myth in terrorist operations: the asymmetrical threat of simple intelligence. Technical report, SANS InfoSec Reading Room. http://www.sans.org/reading_room/whitepapers/stenganography/.

Bas, P., Filler, T. and Pevný, T. 2011. 'Break our steganographic system': the ins and outs of organizing BOSS. In T. Filler, T. Pevný, S. Craver and A.D. Ker (eds), Proceedings of the 13th International Conference on Information Hiding, IH 2011, Prague, Czech Republic, May 18–20, 2011, Revised Selected Papers, vol. 6958 of Lecture Notes in Computer Science. Springer-Verlag.

Batagelj, V. and Bren, M. 1995. Comparing resemblance measures. *Journal of Classification* **12**, 73–90.

BBC News. 2006. Reuters drops Beirut photographer. http://news.bbc.co.uk/2/hi/middle_east/5254838.stm.

Bhat, V.H., Krishna, S., Shenoy, P.D., Venugopal, K.R. and Patnaik, L.M. 2010. JPEG steganalysis using HBCL statistics and FR index In PAISI. In H. Chen, M. Chau, S.H. Li, S.R. Urs, S. Srinivasa and G.A. Wang (eds), *Proceedings of Pacific Asia Workshop*, PAISI 2010, Hyderabad, India, June 21, 2010, vol. 6122 of Lecture Notes in Computer Science. Springer-Verlag, pp. 105–112.

Bhattacharyya, G.K. and Johnson, R.A. 1977. *Statistical Concepts and Methods*. John Wiley & Sons, Inc.

Billingsley, P. 1995. *Probability and Measure*. John Wiley & Sons, Inc.

Böhme, R. 2005. Assessment of steganalytic methods using multiple regression models. In M. Barni, J. Herrera-Joancomartí, S. Katzenbeisser and F. Pérez-González (eds), Proceedings of the 7th International Conference on Information Hiding, vol. 3727 of Lecture Notes in Computer Science. Springer-Verlag, pp. 278–295.

Boser, B.E., Guyon, I. and Vapnik, V. 1992. A training algorithm for optimal margin classifiers. Proceedings of Fifth ACM Workshop on Computational Learning Theory, pp. 144–152.

Briffa, J., Ho, A.T., Schaathun, H.G. and Wahab, A.W.A. 2009a. Conditional probability-based steganalysis for JPEG steganography. International Conference on Signal Processing Systems (ICSPS 2009).

Briffa, J.A., Schaathun, H.G. and Wahab, A.W.A. 2009b. Has F5 really been broken? International Conference on Imaging for Crime Detection and Prevention (ICDP).

Brown, G. 2009. A new perspective for information theoretic feature selection. Twelfth International Conference on Artificial Intelligence and Statistics, Florida.

Burges, C.J.C. 1998. A tutorial on support vector machines for pattern recognition. *Data Mining and Knowledge Discovery* **2**, 121–167.

Cachin, C. 1998. An information-theoretic model for steganography. In M. Barni, J. Herrera-Joancomartí, S. Katzenbeisser and F. Pérez-González (eds), Proceedings of the 1st International Conference on Information Hiding, vol. 3727 of Lecture Notes in Computer Science. Springer-Verlag, pp. 306–318.

Cancelli, G., Doërr, G., Cox, I. and Barni, M. 2008. Detection of ±1 steganography based on the amplitude of histogram local extrema. Proceedings IEEE, International Conference on Image Processing (ICIP).

Chang, C.C. and Lin, C.J. 2000. Training v-support vector classifiers: theory and algorithms. *Neural Computation* **13**(9), 2119–2147.

Chang, C.C. and Lin, C.J. 2001. LIBSVM: a library for support vector machines. Software available at http://www.csie.ntu.edu.tw/~cjlin/libsvm.

Chang, C.C. and Lin, C.J. 2011. LIBSVM FAQ. http://www.csie.ntu.edu.tw/~cjlin/libsvm/faq.html.

Chen, C. and Shi, Y.Q. 2008. JPEG image steganalysis utilizing both intrablock and interblock correlations. ISCAS, May 18–21, 2008, Seattle, WA. IEEE, pp. 3029–3032.

Chen, C., Shi, Y.Q., Chen, W. and Xuan, G. 2006a. Statistical moments based universal steganalysis using JPEG 2-d array and 2-d characteristic function. Proceedings of the International Conference on Image Processing, ICIP 2006, Atlanta, GA, pp. 105–108.

Chen, X., Wang, Y., Tan, T. and Guo, L. 2006b. Blind image steganalysis based on statistical analysis of empirical matrix. Proceedings of 18th International Conference on Pattern Recognition, vol. 3, pp. 1107–1110.

Chupeau, B. 2009. Digital media forensics (part 2). *The Security Newsletter*.

Cohen, G., Honkala, I., Litsyn, S. and Lobstein, A. 1997. *Covering Codes*. Vol. 54 of North-Holland Mathematical Library. North-Holland Publishing Co.

Cortes, C. and Vapnik, V. 1995. Support-vector network. *Machine Learning* **20**, 273–297.

Cox, J., Miller, M. and Bloom, J. 2002. *Digital Watermarking*. Morgan Kaufmann.

Cox, I., Miller, M., Bloom, J., Friedrich, J. and Kalker, T. 2007. *Digital Watermarking and Steganography*, 2nd edn. Morgan Kaufmann.

Crandall, R. 1998. Some notes on steganography. Posted on Steganography Mailing List.

Delp, E.J. and Wong, P.W. (eds). 2004. Security, Steganography, and Watermarking of Multimedia Contents VI, San Jose, CA, January 18–22, 2004, vol. 5306 of SPIE Proceedings. SPIE.

Delp, E.J. and Wong, P.W. (eds). 2005. Security, Steganography, and Watermarking of Multimedia Contents VII, San Jose, CA, January 17–20, 2005, vol. 5681 of SPIE Proceedings. SPIE.

Dong, J. and Tan, T. 2008. Blind image steganalysis based on run-length histogram analysis. 15th IEEE International Conference on Image Processing, pp. 2064–2067.

Dong, J., Chen, X., Guo, L. and Tan, T. 2008. Fusion based blind image steganalysis by boosting feature selection. Proceedings of 6th International Workshop on Digital Watermarking. Springer-Verlag, pp. 87–98.

Dong, J., Wang, W. and Tan, T. 2009. Multi-class blind steganalysis based on image run-length analysis. *Digital Watermarking*, 8th International Workshop, Guildford, August 24–26, 2009, vol. 5703 of Lecture Notes in Computer Science. Springer-Verlag, pp. 199–210.

Dumitrescu, S., Wu, X. and Memon, N. 2002. On steganalysis of random LSB embedding in continuous tone images. Proceedings ICIP, Rochester, NY, September 22–25, 2002, pp. 324–339.

Ecrypt Network of Excellence. 2008. Break our watermarking system [includes the 10 000-image BOWS2 original database intended for information hiding experimentation].

Efron, B. and Tibshirani, R.J. 1993. *An Introduction to the Bootstrap*. Chapman & Hall/CRC.

Esteves, R.M. and Rong, C. 2011. Using Mahout for clustering Wikipedia's latest articles: a comparison between k-means and fuzzy c-means in the cloud. In C. Lambrinoudakis, P. Rizomiliotis and T.W. Wlodarczyk (eds), *CloudCom*. IEEE, pp. 565–569.

Farid, H. n.d. Steganography. Online resource including Matlab source code at http://www.cs.dartmouth.edu/farid/research/steganography.html.

Fawcett, T. 2003. ROC graphs: notes and practical considerations for researchers. Technical report, HP Labs.

Fawcett, T. 2006. An introduction to ROC analysis. *Pattern Recognition Letters* **27**(8), 861–874.

Fawcett, T. and Flach, P.A. 2005. A response to Webb and Ting's 'The application of ROC analysis to predict classification performance under varying class distributions'. *Machine Learning* **58**(1), 33–38.

Filler, T. and Fridrich, J. 2009a. Complete characterization of perfectly secure stego-systems with mutually independent embedding operation. Proceedings of the 2009 IEEE International Conference on Acoustics, Speech and Signal Processing. IEEE, pp. 1429–1432.

Filler, T. and Fridrich, J.J. 2009b. Fisher information determines capacity of secure steganography. In S. Katzenbeisser and A.R. Sadeghi (eds), Proceedings of the 11th International Conference on Information Hiding, vol. 5806 of Lecture Notes in Computer Science. Springer-Verlag, pp. 31–47.

Filler, T., Pevný, T., Craver, S. and Ker, A.D. (eds). 2011a. Proceedings of the 13th International Conference on Information Hiding, Prague, Czech Republic, May 18–20, 2011, Revised Selected Papers, vol. 6958 of Lecture Notes in Computer Science. Springer-Verlag.

Filler, T., Pevný, T. and Bas, P. 2011b. Break our steganographic system. Includes a 10 000-image database intended for information hiding experimentation. http://www.agents.cz/boss/BOSSFinal/.

Fisher, R.A. 1936. The use of multiple measurements in taxonomic problems. *Annals of Human Genetics* 7(2), 179–188.

Fridrich, J. 2005. Feature-based steganalysis for JPEG images and its implications for future design of steganographic schemes. Proceedings of the 7th International Conference on Information Hiding, vol. 3200 of Lecture Notes in Computer Science. Springer-Verlag, pp. 67–81.

Fridrich, J. 2009. *Steganography in Digital Media*. Cambridge University Press.

Fridrich, J. and Long, M. 2000. Steganalysis of LSB encoding in color images. Multimedia and Expo 2000, IEEE International Conference, vol. 3. IEEE, pp. 1279–1282.

Fridrich, J., Goljan, M. and Du, R. 2001a. Steganalysis based on JPEG compatibility. SPIE Multimedia Systems and Applications IV, pp. 275–280.

Fridrich, J., Goljan, M. and Du, R. 2001b. Detecting LSB steganography in color and grayscale images. *IEEE Multimedia and Security* 8(4), 22–28.

Fridrich, J., Goljan, M. and Hogea, D. 2002. Attacking the OutGuess. ACM Workshop on Multimedia and Security, pp. 3–6.

Fridrich, J., Goljan, M., Hogea, D. and Soukal, D. 2003a. Quantitative steganalysis of digital images: estimating the secret message length. *Multimedia Systems* 9(3), 288–302.

Fridrich, J.J., Goljan, M. and Hogea, D. 2003b. Steganalysis of JPEG images: breaking the F5 algorithm. Revised Papers from the 5th International Workshop on Information Hiding. Springer-Verlag, pp. 310–323.

Fridrich, J., Pevný, T. and Kodovský, J. 2007. Statistically undetectable JPEG steganography: dead ends, challenges, and opportunities. Proceedings of the 9th Workshop on Multimedia & Security. ACM, pp. 3–14.

Fridrich, J.J., Kodovský, J., Holub, V. and Goljan, M. 2011a. Breaking HUGO – the process discovery. In Filler, T., Pevný, T., Craver, S. and Ker, A.D. (eds), Proceedings of the 13th International Conference on Information Hiding, Prague, Czech Republic, May 18–20, 2011, Revised Selected Papers, vol. 6958 of Lecture Notes in Computer Science. Springer-Verlag.

Fridrich, J.J., Kodovský, J., Holub, V. and Goljan, M. 2011b. Steganalysis of content-adaptive steganography in spatial domain. In Filler, T., Pevný, T., Craver, S. and Ker, A.D. (eds), Proceedings of the 13th International Conference on Information Hiding, Prague, Czech Republic, May 18–20, 2011, Revised Selected Papers, vol. 6958 of Lecture Notes in Computer Science. Springer-Verlag.

Givner-Forbes, R. 2007. Steganography: Information technology in the service of jihad. Technical report, The International Centre for Political Violence and Terrorism Research, Nanyang Technological University, Singapore.

Goljan, M., Fridrich, J. and Holotyak, T. 2006. New blind steganalysis and its implications. Electronic Imaging, Security, Steganography, and Watermarking of Multimedia Contents VIII, vol. 6072 of SPIE Proceedings. SPIE, pp. 1–13.

Gonzalez, R.C. and Woods, R.E. 2008. *Digital Image Processing*, 3rd edn. Pearson Prentice Hall.

Goth, G. 2005. Steganalysis gets past the hype. *IEEE Distributed Systems Online* 6, 1–5.

Greenspun, P. 2011. Photography. http://philip.greenspun.com/photography/.

Gul, G. and Kurugollu, F. 2011. A new methodology in steganalysis: breaking highly undetectable steganograpy (HUGO). In Filler, T., Pevný, T., Craver, S. and Ker, A.D. (eds), Proceedings of the 13th International Conference on Information Hiding, Prague, Czech Republic, May 18–20, 2011, Revised Selected Papers, vol. 6958 of Lecture Notes in Computer Science. Springer-Verlag.

Guorong, X., Peiqi, C. and Minhui, W. 1996. Bhattacharyya distance feature selection. Proceedings of the 13th International Conference on Pattern Recognition, August 25–29, 1996, Vienna, Austria, vol. 2. IEEE, pp. 195–199.

Hair, J. 2006. *Multivariate Data Analysis*. Pearson Prentice Hall.

Hardy, G.H., Littlewood, J.E. and Pólya, G. 1934. *Inequalities*. Cambridge University Press.

Harmsen, J. 2003. Steganalysis of additive noise modelable information hiding. Master's thesis, Rensselaer Polytechnic Institute.

Harmsen, J. and Pearlman, W. 2003. Steganalysis of additive noise modelable information hiding. SPIE Electronic Imaging.

Harmsen, J.J. and Pearlman, W.A. 2009. Capacity of steganographic channels. *IEEE Transactions on Information Theory* **55**(4), 1775–1792.

Harmsen, J.J., Bowers, K.D. and Pearlman, W.A. 2004. Fast additive noise steganalysis. Security, Steganography, and Watermarking of Multimedia Contents VI, San Jose, CA, January 18–22, 2004, vol. 5306 of SPIE Proceedings. SPIE.

Herodotus. 440 BC. *The Histories*. http://classics.mit.edu/Herodotus/history.5.v.html.

Hofbauer, K., Kubin, G. and Kleijn, W.B. 2009. Speech watermarking for analog flat-fading bandpass channels. *Transactions on Audio, Speech and Language Processing* **17**, 1624–1637.

Holotyak, T., Fridrich, J. and Voloshynovskiy, S. 2005. Blind statistical steganalysis of additive steganography using wavelet higher order statistics. 9th IFIP TC-6 TC-11 Conference on Communications and Multimedia Security, vol. 3677 of Lecture Notes in Computer Science. Springer-Verlag, pp. 273–274.

Hopper, N.J. 2004. Toward a theory of steganography. PhD thesis, School of Computer Science, Carnegie Mellon University.

Hsu, C.W. and Lin, C.J. 2002. A comparison of methods for multi-class support vector machines. *IEEE Transactions on Neural Networks* **13**, 415–425.

Hsu, C.W., Chang, C.C. and Lin, C.J. 2003. A practical guide to support vector classification. Technical report, Department of Computer Science, National Taiwan University.

Hunter, J., Dale, D. and Droettboom, M. 2011. Matplotlib documentation. http://matplotlib.sourceforge.net/contents.html.

IJG. 1998. libjpeg. The Independent JPEG Group's Implementation of the JPEG standard; available at http://www.ijg.org/.

Jenssen, R. and Eltoft, T. 2006. An information theoretic perspective to kernel *k*-means. Proceedings of IEEE International Workshop on Machine Learning for Signal Processing, pp. 161–166.

Jeruchim, M.C., Balaban, P. and Shanmugan, K.S. 2000. *Simulation of Communication Systems. Modeling, Methodology, and Techniques*, 2nd edn. Kluwer Academic/Plenum.

Johnson, R.A. and Wichern, D.W. 1988. *Applied Multivariate Statistical Analysis*, 2nd edn. Prentice Hall.

Kelley, J. 2001. Terrorist instruction hidden online. *USA Today*.

Ker, A.D. 2004a. Improved detection of LSB steganography in grayscale images. Proceedings of the 6th Information Hiding Workshop, vol. 3200 of Lecture Notes in Computer Science. Springer-Verlag, pp. 97–115.

Ker, A.D. 2004b. Quantitative evaluation of pairs and RS steganalysis. Security, Steganography, and Watermarking of Multimedia Contents VI, San Jose, CA, January 18–22, 2004, vol. 5306 of SPIE Proceedings. SPIE.

Ker, A.D. 2005a. Resampling and the detection of LSB matching in colour bitmaps. Security, Steganography, and Watermarking of Multimedia Contents VII, San Jose, CA, January 17–20, 2005, vol. 5681 of SPIE Proceedings. SPIE.

Ker, A.D. 2005b. Steganalysis of LSB matching in grayscale images. *Signal Processing Letters* **12**(6), 441–444.

Ker, A.D. 2006. Batch steganography and pooled steganalysis. In J. Camenisch, C.S. Collberg, N.F. Johnson and P. Sallee (eds), Proceedings of the 8th International Conference on Information Hiding, vol. 4437 of Lecture Notes in Computer Science. Springer-Verlag, pp. 265–281.

Ker, A.D. 2007a. Batch steganography and the threshold game. Security, Steganography, and Watermarking of Multimedia Contents IX, vol. 6505 of SPIE Proceedings. SPIE, pp. 401–413.

Ker, A.D. 2007b. A capacity result for batch steganography. *IEEE Signal Processing Letters* **14**(8), 525–528.

Ker, A.D. 2010a. The square root law does not require a linear key. Proceedings of the 12th ACM Workshop on Multimedia and Security. ACM, pp. 213–224.

Ker, A.D. 2010b. The square root law in stegosystems with imperfect information. In R. Böhme, P.W.L. Fong and R. Safavi-Naini (eds), Proceedings of the 12th International Conference on Information Hiding, vol. 6387 of Lecture Notes in Computer Science. Springer-Verlag, pp. 145–160.

Ker, A.D., Pevný, T., Kodovský, J. and Fridrich, J. 2008. The square root law of steganographic capacity. 10th Multimedia and Security Workshop, ACM Proceedings. ACM, pp. 107–116.

Kerckhoffs, A. 1883. La cryptographie militaire. *Journal des sciences militaires* **IX**, 5–38.

Kharrazi, M., Sencar, H.T. and Memon, N.D. 2005. Benchmarking steganographic and steganalysis techniques. Security, Steganography, and Watermarking of Multimedia Contents VII, San Jose, CA, January 17–20, 2005, vol. 5681 of SPIE Proceedings. SPIE.

Kharrazi, M., Sencar, H.T. and Memon, N. 2006a. Cover selection for steganographic embedding. IEEE International Conference on Image Processing.

Kharrazi, M., Sencar, H.T. and Memon, N. 2006b. Performance study of common image steganography and steganalysis techniques. *Journal of Electronic Imaging* **15**(4), 041104.

Kharrazi, M., Sencar, H.T. and Memon, N.D. 2006c. Improving steganalysis by fusion techniques: a case study with image steganography. *Transactions on DHMS I*, LNCS 4300, pp. 123–137. Springer-Verlag.

Kim, Y., Duric, Z. and Richards, D. 2007. Modified matrix encoding technique for minimal distortion steganography. Proceedings of the 9th International Conference on Information Hiding, vol. 4437 of Lecture Notes in Computer Science. Springer-Verlag, pp. 314–327.

Kittler, J., Hatef, M., Duin, R. and Matas, J. 1998. On combining classifiers. *IEEE Transactions on Pattern Analysis and Machine Intelligence* **20**(3), 226–239.

Knerr, S., Personnaz, L. and Dreyfus, G. 1990. Single layer learning revisited: a stepwise procedure for building and training a neural network. In J. Fogelman (ed.), *Neurocomputing: Algorithms, Architectures and Applications*. Springer-Verlag.

Kodovský, J. and Fridrich, J.J. 2008. On completeness of feature spaces in blind steganalysis. Proceedings of the 10th ACM Workshop on Multimedia and Security. ACM, pp. 123–132.

Kodovský, J. and Fridrich, J. 2009. Calibration revisited. Proceedings of the 11th ACM Workshop on Multimedia and Security. ACM, pp. 63–74.

Kodovský, J. and Fridrich, J. 2011. Feature extractors for steganalysis (web page).

Kohonen, T. 1990. The self-organizing map. Proceedings of the IEEE.

Kraskov, A., Stögbauer, H. and Grassberger, P. 2004. Estimating mutual information. *Phys. Rev. E* **69**, 066138.

Lam, E.Y. and Goodman, J.W. 2000. A mathematical analysis of the DCT coefficient distributions for images. *IEEE Transactions on Image Processing* **9**(10), 1661–1666.

Lee, K., Westfeld, A. and Lee, S. 2006. Category attack for LSB steganalysis of JPEG images. In Y. Shi and B. Jeon (eds), *Digital Watermarking*, vol. 4283 of Lecture Notes in Computer Science. Springer-Verlag, pp. 35–48.

Lee, K., Westfeld, A. and Lee, S. 2007. Generalised category attack – improving histogram-based attack on JPEG LSB embedding. In T. Furon, F. Cayre, G.J. Doërr and P. Bas (eds), Proceedings of the 9th International Conference on Information Hiding, vol. 4567 of Lecture Notes in Computer Science. Springer-Verlag, pp. 378–391.

Li, X., Zeng, T. and Yang, B. 2008a. Detecting LSB matching by applying calibration technique for difference image. Proceedings of the 10th ACM Multimedia & Security Workshop, pp. 133–138.

Li, X., Zeng, T. and Yang, B. 2008b. A further study on steganalysis of LSB matching by calibration. 15th IEEE International Conference on Image Processing (ICIP 2008), pp. 2072–2075.

Lie, W. and Lin, G. 2005. A feature-based classification technique for blind image steganalysis. *IEEE Transactions on Multimedia* **7**(6), 1007–1020.

Lin, G.S., Yeh, C. and Kuo, C.C. 2004. Data hiding domain classification for blind image steganalysis. IEEE International Conference on Multimedia and Expo, vol. 2, pp. 907–910.

Liu, Q., Sung, A.H., Chen, Z. and Xu, J. 2008. Feature mining and pattern classification for steganalysis of LSB matching steganography in grayscale images. *Pattern Recognition* **41**(1), 56–66.

Luger, G.F. 2008. *Artificial Intelligence: Structures and Strategies for Complex Problem Solving*, 6th edn. Addison-Wesley.

Lutz, M. 2010. *Programming Python*, 4th edn. O'Reilly Media.

Lyu, S. and Farid, H. 2003. Detecting hidden messages using higher-order statistics and support vector machines. Revised Papers from the 5th International Workshop on Information Hiding, Lecture Notes in Computer Science. Springer-Verlag, pp. 340–354.

Lyu, S. and Farid, H. 2004. Steganalysis using color wavelet statistics and one-class support vector machines. Security, Steganography, and Watermarking of Multimedia Contents VI, San Jose, CA, January 18–22, 2004, vol. 5306 of SPIE Proceedings. SPIE.

Lyu, S. and Farid, H. 2005. How realistic is photorealistic? *IEEE Transactions on Signal Processing* **53**(2), 845–850.

Lyu, S. and Farid, H. 2006. Steganalysis using higher-order image statistics. *IEEE Transactions on Information Forensics and Security* **1**(1), 111–119.

MacKay, D.J. 2003. *Information Theory, Inference, and Learning Algorithms*. Cambridge University Press.

Macskassy, S.A. and Provost, F.J. 2004. Confidence bands for ROC curves: methods and an empirical study. In J. Hernández-Orallo, C. Ferri, N. Lachiche and P.A. Flach (eds), Proceedings of the First Workshop on ROC Analysis in AI, August 2004, pp. 61–70.

Mallat, S. 1989a. Multiresolution approximation and wavelet orthonormal bases of l^2. *Transactions of the American Mathematical Society* **315**, 69–87.

Mallat, S.G. 1989b. A theory for multiresolution signal decomposition: the wavelet representation. *IEEE Transactions on Pattern Analysis and Machine Intelligence* **11**(7), 674–693.

Marsland, S. 2009. *Machine Learning. An Algorithmic Perspective*. CRC Press.

Marvel, L.M., Boncelet Jr., C.G. and Retter, C.T. 1999. Spread spectrum image steganography. *IEEE Transactions on Image Processing* **8**(8), 1075–1083.

Maurer, U.M. 1992. A universal statistical test for random bit generators. *Journal of Cryptology* **5**, 89–105.

McGill, W.J. 1954. Multivariate information transmission. *IEEE Transactions on Information Theory* **4**(4), 93–111.

Miche, Y., Roue, B., Lendasse, A. and Bas, P. 2006. A feature selection methodology for steganalysis. In B. Günsel, A.K. Jain, A.M. Tekalp and B. Sankur (eds), MRCS, vol. 4105 of Lecture Notes in Computer Science. Springer-Verlag, pp. 49–56.

Miche, Y., Bas, P., Lendasse, A., Jutten, C. and Simula, O. 2007. Advantages of using feature selection techniques on steganalysis schemes. In F.S. Hernández, A. Prieto, J. Cabestany and M. Gra na (eds), IWANN, vol. 4507 of Lecture Notes in Computer Science. Springer-Verlag, pp. 606–613.

Miche, Y., Bas, P., Lendasse, A., Jutten, C. and Simula, O. 2009. Reliable steganalysis using a minimum set of samples and features. *EURASIP Journal on Information Security* **2009**, pp. 1–13.

Mihçak, M.K., Kozintsev, I. and Ramchandran, K. 1999a. Spatially adaptive statistical modeling of wavelet image coefficients and its application to denoising. Proceedings of IEEE International Conference on Acoustics, Speech and Signal Processing.

Mihçak, M.K., Kozintsev, I., Ramchandran, K. and Moulin, P. 1999b. Low complexity image denoising based on statistical modeling of wavelet coefficients. *IEEE Signal Processing Letters* **6**(12), 300–303.

Murty, M. and Devi, V. 2011. *Pattern Recognition: An Algorithmic Approach*. Undergraduate Topics in Computer Science. Springer-Verlag.

Negnevitsky, M. 2005. *Artificial Intelligence. A Guide to Intelligent Systems*, 2nd edn. Addison-Wesley.

Nelson, D. (ed.). 1998. *Dictionary of Mathematics*, 2nd edn. Penguin.

Nill, N. 1985. A visual model weighted cosine transform for image compression and quality assessment. *IEEE Transactions on Communications* **33**(6), 551–557.

Ojala, T., Pietikäinen, M. and Harwood, D. 1996. A comparative study of texture measures with classification based on featured distributions. *Pattern Recognition* **29**(1), 51–59.

Parzen, E. 1962. On estimation of a probability density function and mode. *Annals of Mathematics & Statistics* **33**(3), 1065–1076.

Petitcolas, F. 2009. Photo database. http://www.petitcolas.net/fabien/watermarking/image_database/.

Pevný, T. and Fridrich, J. 2007. Merging Markov and DCT features for multi-class JPEG steganalysis. Proceedings of SPIE Electronic Imaging. SPIE, pp. 3–4.

Pevný, T. and Fridrich, J. 2008. Multiclass detector of current steganographic methods for JPEG format. *IEEE Transactions on Information Forensics and Security* **3**(4), 635–650.

Pevný, T., Bas, P. and Fridrich, J. 2009a. Steganalysis by subtractive pixel adjacency matrix. Proceedings of the 11th ACM Workshop on Multimedia and Security. ACM, pp. 75–84.

Pevný, T., Fridrich, J. and Ker, A.D. 2009b. From blind to quantitative steganalysis. In Media Forensics and Security XI, vol. 7254 of SPIE Proceedings. SPIE, pp. C01–C14.

Pevný, T., Filler, T. and Bas, P. 2010a. Using high-dimensional image models to perform highly undetectable steganography. Proceedings of the 12th International Conference on Information Hiding. ACM, pp. 161–177.

Pevný, T., Bas, P. and Fridrich, J.J. 2010b. Steganalysis by subtractive pixel adjacency matrix. *IEEE Transactions on Information Forensics and Security* **5**(2), 215–224.

Provos, N. 2004. Outguess. http://www.outguess.org/.

Provos, N. and Honeyman, P. 2001. Detecting steganographic content on the Internet. Technical Report 01-11, Center for Information Technology Integration, University of Michigan.

Provos, N. and Honeyman, P. 2003. Hide and seek: an introduction to steganography. *IEEE Security and Privacy* **1**(3), 32–44.

Pudil, P., Novovicová, J. and Kittler, J. 1994. Floating search methods in feature selection. *Pattern Recognition Letters* **15**, 1119–1125.

Rocha, A. 2006. Randomizaç ao progressiva para esteganálise. Master's thesis, Instituto de Computaç ao – Unicamp, Campinas, SP, Brasil.

Rocha, A. and Goldenstein, S. 2006a. Progressive randomization for steganalysis. 8th IEEE International Conference on Multimedia and Signal Processing.

Rocha, A. and Goldenstein, S. 2006b. Progressive randomization for steganalysis. Technical Report IC-06-07, Universidade Estadual de Campinas.

Rocha, A. and Goldenstein, S. 2008 Steganography and steganalysis in digital multimedia: hype or hallelujah? *RITA* **XV**(1), 83–110.

Rocha, A. and Goldenstein, S. 2010. Progressive randomization: seeing the unseen. *Computer Vision and Image Understanding* **114**(3), 349–362.

Rosenblatt, M. 1956. Remarks on some nonparametric estimates of a density function. *Annals of Mathematics & Statistics* **27**, 832–837.

Sajedi, H. and Jamzad, M. 2009. Secure cover selection steganography. In J.H. Park, H.H. Chen, M. Atiquzzaman, C. Lee, T.H. Kim and S.S. Yeo (eds), ISA, vol. 5576 of Lecture Notes in Computer Science. Springer-Verlag, pp. 317–326.

Sallee, P. 2009. JPEG Toolbox for Matlab. Software available for free download.

Schaathun, H.G. 2011. Vurdering av features for steganalyse i JPEG. Norsk Informatikkonferanse, Tromsø, Norway.

Schaefer, G. and Stich, M. 2004. UCID – an uncompressed colour image database. Proceedings of SPIE, Storage and Retrieval Methods and Applications for Multimedia 2004. SPIE, pp. 472–480.

Schölkopf, B., Smola, A.J., Williamson, R.C. and Bartlett, P.L. 2000. New support vector algorithms. *Neural Computing* **12**, 1207–1245.

Schölkopf, B., Platt, J.C., Shawe-Taylor, J.C., Smola, A.J. and Williamson, R.C. 2001. Estimating the support of a high-dimensional distribution. *Neural Computing* **13**, 1443–1471.

Scott, D.W. 2009. Sturges' rule. *Wiley Interdisciplinary Reviews: Computational Statistics* **1**(3), 303–306.

Service NRC. 2009. NRCS photo gallery. Image database.

Shannon, C.E. 1948. A mathematical theory of communication. *The Bell System Technical Journal* **27**, 379–423.

Shi, Y.Q. 2009. Research on steganalysis. http://web.njit.edu/~shi/Steganalysis/steg.htm.

Shi, Y., Xuan, G., Zou, D., Gao, J., Yang, C., Zhang, Z., Chai, P., Chen, W. and Chen, C. 2005a. Image steganalysis based on moments of characteristic functions using wavelet decomposition, prediction error image, and neural network. IEEE International Conference on Multimedia and Expo 2005, ICME 2005. IEEE, 4 pp.

Shi, Y.Q., Xuan, G., Yang, C., Gao, J., Zhang, Z., Chai, P., Zou, D., Chen, C. and Chen, W. 2005b. Effective steganalysis based on statistical moments of wavelet characteristic function. ITCC (1), pp. 768–773. IEEE Computer Society.

Shi, Y.Q., Chen, C. and Chen, W. 2006. A Markov process based approach to effective attacking JPEG steganography. Proceedings of 8th International Conference on Information Hiding, Lecture Notes in Computer Science. Springer-Verlag.

Simmons, G.J. 1983. The prisoners' problem and the subliminal channel. CRYPTO, pp. 51–67.

Simoncelli, E.P. and Adelson, E.H. 1990. Subband transforms. In J.W. Woods (ed.), *Subband Image Coding*. Kluwer Academic, pp. 143–192.

Spackman, K. 1989. Signal detection theory: valuable tools for evaluating inductive learning. Proceedings of Sixth International Workshop on Machine Learning. Morgan-Kaufman, pp. 160–163.

Sullivan, K., Madhow, U., Chandrasekaran, S. and Manjunath, B.S. 2005. Steganalysis of spread spectrum data hiding exploiting cover memory. Security, Steganography, and Watermarking of Multimedia Contents VII, San Jose, CA, January 17–20, 2005, vol. 5681 of SPIE Proceedings. SPIE.

Swets, J. 1988. Measuring the accuracy of diagnostic systems. *Science* **240**, 1285–1293.

Theodoridis, S. and Koutroumbas, K. 2009. *Pattern Recognition*, 4th edn. Academic Press.

Tieu, K. and Viola, P. 2004. Boosting image retrieval. *International Journal of Computer Vision* **56**(1/2), 17–36.

Trithemius, J. c. 1500. Steganographie: Ars per occultam Scripturam animi sui voluntatem absentibus aperiendi certu. Three volumes. First printed in Frankfurt, 1606.

Unknown. 2007. Secret information: hide secrets inside of pictures. *The Technical Mujahedin* **2**, 1–18 (in Arabic).

Vapnik, V.N. 2000. *The Nature of Statistical Learning Theory*, 2nd edn. Springer-Verlag.

Vapnik, V. and Lerner, A. 1963. Pattern recognition using generalized portrait method. *Automation and Remote Control* **24**, 774–780.

Wahab, A.W.A. 2011. Image steganalysis and machine learning. PhD thesis, University of Surrey.

Wahab, A.W.A., Schaathun, H.G. and Ho, A.T. 2009. Markov process based steganalysis by using second-order transition probability matrix. 8th European Conference on Information Warfare and Security, ECIW 09, Military Academy, Lisbon and the University of Minho, Braga, Portugal.

Wang, L. and He, D.C. 1990. Texture classification using texture spectrum. *Pattern Recognition* **23**, 905–910.

Wayner, P. 2002. *Disappearing Cryptography*, 2nd edn. Morgan Kaufmann.

Weinberger, M., Seroussi, G. and Sapiro, G. 1996. Locoi: a low complexity context-based loss-less image compression algorithm. Proceedings of IEEE Data Compression Conference, pp. 140–149.

Westfeld, A. 2001. F5 – a steganographic algorithm. Proceedings of the 4th International Workshop on Information Hiding. Springer-Verlag, pp. 289–302.

Westfeld, A. 2003. Detecting low embedding rates. Revised Papers from the 5th International Workshop on Information Hiding. Springer-Verlag, pp. 324–339.

Westfeld, A. 2007. ROC curves for steganalysts. Proceedings of the Third WAVILA Challenge (Wacha '07), pp. 39–45.

Westfeld, A. and Pfitzmann, A. 2000. Attacks on steganographic systems. In A. Pfitzmann (ed.), *Information Hiding*, vol. 1768 of Lecture Notes in Computer Science. Springer-Verlag, pp. 61–75.

Xuan, G., Gao, J., Shi, Y. and Zou, D. 2005a. Image steganalysis based on statistical moments of wavelet subband histograms in DFT domain. 7th IEEE Workshop on Multimedia Signal Processing. IEEE, pp. 1–4.

Xuan, G., Shi, Y.Q., Gao, J., Zou, D., Yang, C., Zhang, Z., Chai, P., Chen, C. and Chen, W. 2005b. Steganalysis based on multiple features formed by statistical moments of wavelet character-istic functions. In M. Barni, J. Herrera-Joancomartí, S. Katzenbeisser and F. Pérez-González (eds), *Information Hiding*, vol. 3727 of Lecture Notes in Computer Science. Springer-Verlag, pp. 262–277.

Yadollahpour, A. and Naimi, H.M. 2009. Attack on LSB steganography in color and grayscale images using autocorrelation coefficients. *European Journal of Scientific Research* **31**(2), 172–183.

Yang, H.H. and Moody, J. 1999. Data visualization and feature selection: new algorithms for non-Gaussian data. In *Advances in Neural Information Processing Systems*. MIT Press, pp. 687–693.

Yang, X.Y., Liu, J., Zhang, M.Q. and Niu, K. 2007. A new multi-class SVM algorithm based on one-class SVM. In Y. Shi, G. van Albada, J. Dongarra and P. Sloot (eds), Computational Science – ICCS 2007, vol. 4489 of Lecture Notes in Computer Science. Springer-Verlag, pp. 677–684.

Yeung, R.W. 2002. A first course in information theory. *Information Technology: Transmission, Processing and Storage*. Kluwer Academic/Plenum Publishers.

Zaker, N. and Hamzeh, A. 2011. A novel steganalysis for TPVD steganographic method based on differences of pixel difference histogram. Multimedia Tools and Applications, pp. 1–20.

Zhang, S. and Zhang, H. 2010. JPEG steganalysis using estimated image and Markov model. In D.S. Huang, T.M. McGinnity, L. Heutte and X.P. Zhang (eds), Advanced Intelligent Computing Theories and Applications – 6th International Conference on Intelligent Computing (ICIC), vol. 93 of Communications in Computer and Information Science. Springer-Verlag, pp. 281–287.

Zhang, J., Cox, I.J. and Doerr, G. 2007. Steganalysis for LSB matching in images with high-frequency noise. IEEE International Workshop on Multimedia Signal Processing (MMSP).

Index

Machine Learning in Image Steganalysis, First Edition. Hans Georg Schaathun.
© 2012 John Wiley & Sons, Ltd. Published 2012 by John Wiley & Sons, Ltd.